New Perspectives on Food Blanching

Felipe Richter Reis

Editor

New Perspectives on Food Blanching

 Springer

Editor
Felipe Richter Reis
Food Technician Course
Instituto Federal do Paraná, Campus
  Jacarezinho
Jacarezinho, Paraná
Brazil

ISBN 978-3-319-83989-9   ISBN 978-3-319-48665-9   (eBook)
DOI 10.1007/978-3-319-48665-9

Printed on acid-free paper

This Springer imprint is published by Springer Nature
The registered company is Springer International Publishing AG
The registered company address is: Gewerbestrasse 11, 6330 Cham, Switzerland

I would like to dedicate this book:

To my wife Daiany, for accepting to begin a new life with me in Jacarezinho,

To my family, especially my mother Margaret, an outstanding woman, and my father Celso, who completed 70 years in 2016,

And to all my students and friends from the IFPR, where this work was developed.

Preface

Blanching is a widely used method in food processing companies, especially for improving the quality of vegetables prior to canning, freezing, drying, and other suitable processing techniques. The positive impact of blanching on food quality is related, for example, to the activation or inactivation of enzymes, decrease of the microbial load and of the content of undesirable substances, and improvement of the extractability of bioactive compounds, thus resulting in improved sensory and nutritional quality. In addition, there are numerous reports on the favorable effect of blanching on the performance of drying. Although most of the reports on blanching highlight its positive effect, reports showing the unfavorable influence of blanching are also available. This book summarizes the results of studies on food blanching performed especially over the past ten years, although older reference studies were also used. Chapter 1 introduces the theme by presenting key concepts, along with previously published documents dealing with blanching from a conceptual viewpoint. Chapter 2 presents the advances on the effect of blanching on food's physical, chemical, and sensory qualities. Once potatoes were the most commonly blanched food before processing, this chapter brings a specific section on potato blanching. Chapter 3 is devoted to a recent topic, namely bioactive compounds, as affected by blanching. In this sense, functional foods are under increasing investigation and their bioactive compounds should be preserved as much as possible during processing, which includes blanching. On the other hand, some compounds are deleterious for humans and blanching could help removing them from food. High microbial loads are also undesirable since they indicate poor hygiene conditions and, depending on the microorganism found, they comprise a microbiological hazard. The decrease in antinutrients, pesticides, and microorganisms is the topic of Chap. 4. A recent health concern is related to the presence of a specific compound in thermally treated food, named acrylamide. Chapter 5 is dedicated to the mitigation of acrylamide by means of blanching. The contribution of drying techniques for food preservation is undoubted. Numerous reports point out to the contribution of blanching for the performance of subsequent drying. For this reason, Chap. 6 is dedicated to this issue. Finally, it is mentioned that blanching sometimes exerts undesirable effects on food quality. Such drawbacks may be overcome by using improved blanching techniques, such as microwave blanching, or by using alternative techniques, such as ultrasound. Therefore, Chap. 7 is dedicated to advanced blanching techniques and alternative pretreatments for improving the food quality. To the best of our knowledge, this is the first book totally devoted to food blanching. I hope that the comprehensive information on blanching brought to the reader in a carefully prepared fashion fulfills the reader's needs.

Jacarezinho, Brazil   Felipe Richter Reis

Contents

Editor and Contributors

About the Editor

Felipe Richter Reis expertise is in fruit and vegetable processing especially by using blanching and drying. He earned his academic degrees in renowned Brazilian universities, and since 2011 he has been professor of several courses in food science and technology. He is constantly conducting research, acting as a reviewer of relevant journals, and publishing scientific material in food science and technology.

Contributors

Bogdan Demczuk Junior is Professor at the Food Engineering undergraduate course of the Federal Technological University of Paraná, where he also conducts research on various subjects, especially on plant foods. His doctorate was focused on annatto seed thereby allowing him to specialize himself on carotenoids.

João Luiz Andreotti Dagostin is a Food Engineer, DSc, with more than ten years research experience in food science and technology. He is currently a postdoctoral fellow at the Federal University of Paraná. In addition, he is associate editor of Boletim do CEPPA, a journal dealing with food science and technology published by the Federal University of Paraná.

Chapter 1
Introduction

Felipe Richter Reis

Abstract Blanching is a suitable way for aiding in the shelf-life increase of various foods prior to other processes of the food industry. Usually it is a process used for vegetables, although there are also reports on the blanching of fruit, animal source foods, fungi, etc. After a search in a renowned database, hundreds of documents dealing with the theme blanching were found. Among them, the few reviews on the theme are presented in this chapter, together with other relevant theoretical considerations. Hot water blanching is the most common process followed by steam blanching. Blanching inactivates inconvenient enzymes that, if active, cause color, texture and flavor impairment. This chapter aims at introducing the theme by providing background information on the issue. It is useful being aware of the principal features of this processing method, which is almost ubiquitous in vegetable processing lines. Though short, it is a valuable chapter for stimulating the reader to continue to the other chapters that go deeper into the new perspectives on the blanching process and blanched foods.

Keywords Blanching · Hot water blanching · Steam blanching · Enzyme inactivation

The practice of storing food has been in place since ancient times in order to avoid famine and for convenience reasons. Once raw food is usually unstable during long-term storage, techniques for increasing food shelf-life have been developed which ultimately led to the emergence of food science and technology.

The use of heat for treating food is also an ancient method. The discovery of fire allowed the use of cooking methods. Nevertheless, the use of heat with goals other than cooking was probably first documented by Appert (1810) in his book *L'art de Conserver Pendant Plusieurs Années toutes les Substances Animales et Végétales.* This document describes a heat treatment of bottled food in boiling water aiming at

F. Richter Reis (✉)
Food Technician Course, Instituto Federal do Paraná, Campus Jacarezinho, Jacarezinho, Paraná, Brazil
e-mail: felipe.reis@ifpr.edu.br

© Springer International Publishing AG 2017
F. Richter Reis (ed.), *New Perspectives on Food Blanching,*
DOI 10.1007/978-3-319-48665-9_1

increasing food shelf-life. Nowadays, the heat treatments used more extensively in the food industry are pasteurization, sterilization and blanching.

Blanching can be defined as a heat treatment aimed at denaturing enzymes in solid fruit and vegetables pieces commonly prior to freezing, drying or canning. Blanching is particularly used for preserving the food color, being usually performed at atmospheric pressure using steam or hot water at temperatures ranging from 90 to 95 °C for around 1 or 2 min (Smith 2011). The above-mentioned conditions comprise the so-called high temperature short time (HTST) blanching. Nevertheless, the use of blanching temperatures as low as 50 °C (Reis et al. 2008) and blanching times as high as 40 min (Fernández et al. 2006) has been reported over the last ten years. After blanching, the product is usually cooled in water for inhibiting thermophiles and avoiding a cooked taste in the product.

Even though inactivating enzymes has been the principal purpose of blanching, reports dating back to the middle of the last century refer to its positive effect in other aspects, such as: (1) reduction of microbial level on foods before freezing (Larkin et al. 1955); (2) improvement on carotenoids retention in dried fruits (Sabry 1961); and (3) reduction of pesticide concentration on vegetable surface (Kleinschmidt 1971; Solar 1971). On the other hand, deleterious effects of blanching have also been reported in that time, namely the loss of water-soluble vitamins in the blanching water (Cook et al. 1961).

The Encyclopedia of Agricultural, Food, and Biological Engineering brings some valuable information in its article named "Blanching of Foods" written by Powers et al. (2004). For these authors, besides inactivating enzymes blanching is also aimed at modifying texture, preserving color, flavor and nutritional value, and removing trapped air. They also mention that color and texture are commonly monitored during blanching processes. In time: the activity of some enzymes is strictly related to color and texture changes in the foods where they are active. For example, polyphenol oxidase (PPO) (EC: 1.10.3.1) catalyzes the enzymatic browning in wounded fruit and vegetables, i.e., the reaction between phenols from the tissue and atmospheric oxygen to produce quinones that polymerize into melanin (Marshall et al. 2000; Severini et al. 2003). Additionally, polygalacturonase (EC: 3.2.1.15) hydrolyzes pectic acid leading to significant decreases in the texture of raw food materials like tomatoes (Whitaker 1996).

Xin et al. (2015) wrote a review on blanching and freezing technologies as applied in plant foods. Among the subjects dealt with were conventional and novel blanching technologies and their effect on product quality. For example, the impact of microwave blanching in nutrient retention, color preservation, enzyme inactivation, and microbial control in different foods was mentioned. Nonthermal treatments such as high-pressure processing and ultrasound were referred to as potential replacers of conventional hot water blanching. Their advantages, like minimal thermal damage, homogeneity of treatment and low energy requirement, were highlighted.

At the domestic level, blanching is usually performed with hot water in a pan. Heat is provided by the flame of the stove and temperature control is deficient. Therefore, domestic blanching is commonly carried out at water boiling temperature that varies according to the local altitude.

Fig. 1.1 An industrial hot water blancher (EMA Europe 2015, used with permission)

Figure 1.1 presents an example of equipment for industrial hot water blanching. The design of the blancher consists of a cylindrical cooking chamber, equipped with a heating jacket and an insulation layer. Inside the chamber, there is a screw with large diameter half immersed in the working vessel. A lifting sectional cover with additional insulation provides convenient access to the entire inner surface of the cooker (EMA Europe 2015). Additionally to the blanchers that use a screw, there are other types of blancher such as the ones that use a conveyor chain to transport the product inside the tank as well as blanchers that are a rotary drum where the product is immersed and conveyed (Powers et al. 2004).

Since vegetables differ in shape, size, heat conductivity, and enzymatic activity, the blanching treatment must be established based on experiments (Dauthy 1995). Data obtained for some vegetables are presented in Table 1.1.

When comparing hot water blanching with steam blanching at industrial level, the latter seems to be more advantageous than the former due to reduced leaching of nutrients and reduced generation of pollutant effluents. Additionally, Wang (2008) states that all blanchers are energy intensive but steam blanchers require from 75 to 90% less energy than hot water blanchers.

The use of steam instead of hot water is only one of the many tentative advances in blanching technology that have been studied over the years. Other studies include: the addition of various types of chemicals to the hot water, the use of microwaves, superheated steam, radio frequency, infrared heating, Ohmic heating and ultrasound. Furthermore, some trials aimed at studying the substitution of blanching by other technologies, such as high-pressure processing, ultraviolet radiation, and high-pressure carbon dioxide.

A search in Scopus database with the terms "blanching and food" led to the download of nearly 350 articles on food blanching from the last ten years. The choice for this period took into account the fact that the book intends mainly to present new perspectives on blanching, once the most basic concepts have been dealt with extensively in previously published documents. Nevertheless, key reference works published before 2006 were used sometimes.

Table 1.1 Blanching conditions for some vegetables

Vegetables	Temperature (°C)	Time (min)
Peas	85–90	2–7
Green beans	90–95	2–5
Cauliflower	Boiling	2
Carrots	90	3–5
Peppers	90	3

Source Dauthy (1995)

This chapter aimed at introducing the most relevant and traditional aspects of food blanching in order to provide basic information on blanching to the reader. In the next chapters, new perspectives on food blanching will be presented thoroughly.

References

Appert N (1810) L'art de Conserver Pendant Plusieurs Années toutes les Substances Animales et Végétales. Patris et Cie, Paris

Cook BB, Gunning B, Uchimoto D (1961) Nutrients in frozen foods: variations in nutritive value of frozen green baby lima beans as a result of methods of processing and cooking. J Agric Food Chem 9:316–321

Dauthy ME (1995) Fruit and vegetable processing. Food and Agriculture Organization of the United Nations, Rome

EMA Europe (2015) NEAEN ThermoScrew Continuous screw cooker/blancher. http://neaen.com/cooking-blanching/neaen-thermoscrew-continuous-screw-cooker-blancher.html. Accessed 26 Nov 2015

Fernández C, Dolores Alvarez M, Canet W (2006) The effect of low-temperature blanching on the quality of fresh and frozen/thawed mashed potatoes. Int J Food Sci Technol 41:577–595. Doi:10.1111/j.1365-2621.2005.01119.x

Kleinschmidt MG (1971) Fate of Di-syston O, O-diethyl S-[2-(Ethylthio)ethyl] phosphorodithioate in potatoes during processing. J Agr Food Chem 19:1196–1197

Larkin EP, Litsky W, Fuller JE (1955) Fecal streptococci in frozen foods. IV. Effect of sanitizing agents and blanching temperatures on *Streptococcus faecalis*. Appl Microbiol 3:107–110

Marshall MR, Kim J, Wei CI (2000) Enzymatic browning in fruits, vegetables and seafoods. Food and Agriculture Organization of the United Nations. http://www.fao.org/ag/ags/agsi/ENZYMEFINALEnzymatic%20Browning.html. Accessed 13 Dec 2005

Powers JR, De Corcuera JIR, Cavalieri RP (2004) Blanching of foods. Encycl Agric Food Biol Eng 1–5. Doi:10.1081/E-EAFE-120030417

Reis FR, Masson ML, Waszczynskyj N (2008) Influence of a blanching pretreatment on color, oil uptake and water activity of potato sticks, and its optimization. J Food Process Eng 31:833–852. Doi:10.1111/j.1745-4530.2007.00193.x

Sabry ZI (1961) Food discoloration, browning in dried fruit products: non-enzymatic browning and its effect on the carotenoids in Qamareddeen, a dried apricot pulp. J Agric Food Chem 9:53–55

Severini C, Baiano A, De Pilli T et al (2003) Prevention of enzymatic browning in sliced potatoes by blanching in boiling saline solutions. LWT - Food Sci Technol 36:657–665. Doi:10.1016/S0023-6438(03)00085-9

Smith PG (2011) Introduction to food process engineering, 2nd edn. Springer, Heidelberg

Solar JM (1971) Removal of aldrin, heptachlor epoxide, and endrin from potatoes during processing. J Agric Food Chem 19:1008–1010

Wang L (2008) Energy efficiency and management in food processing facilities. CRC Press, Boca Raton

Whitaker JR (1996) Enzymes. In: Fennema OR (ed) Food chemistry, 3rd edn. Marcel Dekker, New York

Xin Y, Zhang M, Xu B et al (2015) Research trends in selected blanching pretreatments and quick freezing technologies as applied in fruits and vegetables: a review. Int J Refrig 57:11–25. Doi:10.1016/j.ijrefrig.2015.04.015

Chapter 2
Effect of Blanching on Food Physical, Chemical, and Sensory Quality

Felipe Richter Reis

Abstract Besides the well-known positive impact of blanching on color, the benefits of blanching can reach texture and even flavor. These effects are usually related to the activation or inactivation of key enzymes. Furthermore, reports on the positive effect of blanching on physicochemical parameters of foods are numerous. This chapter aims at reporting the effect of blanching on the physical, chemical, and sensory quality of foods. Given the importance of potatoes for mankind and the numerous studies on the blanching of potatoes, the first section of the chapter is devoted to this topic. Subsequently, blanching of miscellaneous foods is dealt with. The results of these studies are usually presented together with hypotheses that try to explain the observed behavior of foods as affected by blanching. Several equations used for modeling changes in color, texture, and enzymatic activity during blanching are presented. This chapter is very helpful for understanding the impact of blanching on food quality from a general perspective.

Keywords Blanching · Color · Texture · Peroxidase · Polyphenol oxidase

Impact of Blanching on the Quality of Potatoes and Related Products

The focus of this section is to provide information on the effect of blanching on the physical, chemical, and sensory quality of potatoes. Works on the reduction of undesirable compounds, such as acrylamide and pesticides, along with the reduction of microbes in potatoes will be presented in Chaps. 4 and 5. Additionally, works on the influence of blanching on the drying kinetics of potatoes will be shown in Chap. 6.

F. Richter Reis (✉)
Food Technician Course, Instituto Federal do Paraná, Campus Jacarezinho, Jacarezinho, Paraná, Brazil
e-mail: felipe.reis@ifpr.edu.br

© Springer International Publishing AG 2017
F. Richter Reis (ed.), *New Perspectives on Food Blanching*,
DOI 10.1007/978-3-319-48665-9_2

Potatoes are processed in many ways by the food industry. In the potato processing plants, a blanching step is almost mandatory. In order to provide useful information for the potato processing industry, studies on the blanching of potatoes have been extensively conducted over the years. The most ancient report on this issue available in Scopus database is the study of Wallerstein et al. (1947). They report the inhibition of color changes in macerated potatoes by blanching. This effect was probably due to the inactivation of the polyphenol oxidase (PPO) enzyme in the tissue.

There are numerous other studies dealing with aspects such as color and texture in potatoes during blanching. The changes in such parameters were sometimes fitted to kinetic models, helping to explain how texture and color change during blanching and consequently helping to design blanching processes. In addition, studies on the positive impact of blanching on other quality aspects of the final product, like decrease in oil absorption and reduction in water activity of potato chips, are available.

Texture change during blanching is an important phenomenon to be understood and it has been under investigation for a long time. One of these studies showed that two-step blanching leads to higher peak force in potatoes when compared to one-step blanching. The former consisted of immersion in water at 70 °C for 10 min followed by immersion in water at 97 °C for 2 min, and the latter consisted only of immersion in water at 97 °C for 2 min (Agblor and Scanlon 1998).

Texture and color of French fries as affected by blanching, drying, and frying were assessed by Agblor and Scanlon (2000). Their results confirmed their previous study in which a two-step blanching consisting of immersion in water at 70 °C for 10 min followed by immersion in water at 97 °C for 2 min improved the color (higher lightness-L^*-value) and the texture (higher peak force and peak deformation) of potatoes. Nevertheless, one-step blanching at 97 °C for 2 min was suggested by them for processors aiming at promoting softening of firmer tubers (e.g., those with high dry matter content) and maintaining good color quality.

Verlinden et al. (2000) found that hot water blanching of potatoes at 55–75 °C affected their texture parameters, as per reducing the maximum force and the rupture force of the samples, and increasing the deformation at maximum force and the rupture deformation. They attributed this behavior the loss of turgor pressure in the tissue during heating. Additionally, they found that blanching/cooling before cooking had a strengthening effect on potatoes, especially when blanching treatments were longer.

Varnalis et al. (2001a) observed that blanching in boiling water for 2 min before drying of potato cubes led to puffing of the product. Their results show that blanching lead to lower penetration force, lower Young's modulus, and higher volume, which were typical characteristics of puffed cubes. They affirmed that blanching may affect not only the surface of the product, but also the rigidity of the structure allowing the vapor inside the cubes to press the walls of the cubes from inside out, thus promoting puffing. A second study on this issue (Varnalis et al. 2001b) showed that blanching reduces the permeability of a partially dried layer probably due to gelatinization of starch thus allowing puffing. Lewicki and Pawlak

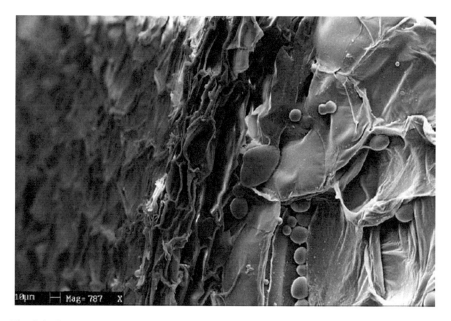

Fig. 2.1 Cross section of a potato tuber with a thin layer of skin and a group of flesh cells with starch granules (Lisińska and Gołubowska 2005)

(2003) confirmed using microscopy that blanching causes gelatinization of potato starch besides increasing the cells size and making their shape more regular.

Lisińska and Gołubowska (2005) studied the structural changes of potato tissue after technological processes used for French fries production, finding that blanching, pre-drying, and frying caused the most important changes in the tissue. When compared to fresh potato strips (Fig 2.1), blanched samples presented starch swelling in the outer layer along with a significant increase in volume (Fig. 2.2). In addition, blanching promoted an increase in non-starch polysaccharides and lignin content, which was related to water losses during the process.

Activation of pectin methyl esterase (PME) enzyme has been associated with better texture in potatoes and related products. For example, a hypothesis on this issue is provided in the study of Aguilar et al. (1997). They attributed firmness improvement in potatoes to demethoxylation caused by PME during blanching at 55–70 °C, which produces free carboxyl groups that can react with ions like calcium and magnesium to yield firmer structure. Aware of the importance and complexity of the impact of PME activity on firmness, Tijskens et al. (1997) modeled the activity of PME enzyme in potatoes as a function of blanching time and blanching temperature. They stated that PME denaturation is achieved in 10 min at 50 °C and in 1 min at 70 °C for potatoes of the cv. Bintje. González-Martínez et al. (2004) studied the temperature distribution in potatoes during blanching at 50–90 °C and its effect on PME activity, finding that heating-induced activation of the enzyme occurs when the

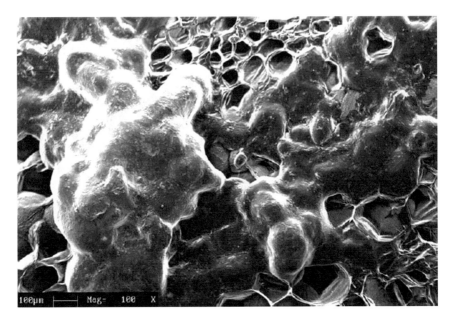

Fig. 2.2 The strips after blanching; starch swelling occurred in the outer layer of strips, along with enormous increase in their volume (Lisińska and Gołubowska 2005)

tissue reaches an average temperature of 52 °C. Another study showed that blanching at 65 °C led to higher firmness in new potatoes (partially grown potatoes) when compared to blanching at 75 °C, being the former the only temperature in which PME remained active even after 30 min of blanching (Abu-Ghannam and Crowley 2006). Therefore, it can be affirmed that PME was activated by blanching at 65 °C exerting a strengthening effect on new potatoes. Carbonell et al. (2006) found that a two-step blanching consisting of immersion in water containing 0.07 mg L^{-1} calcium chloride at 70 °C during 10 min followed by immersion in the same solution at 97 °C for 2 min was an effective pretreatment for preventing firmness loss and inactivating peroxidase (POD) in potatoes after freezing/thawing. This result was also attributed to the activation of the PME enzyme during the first blanching step. The use of blanching before cooking of mashed potatoes led to a desirable light-colored and thickened product, as expressed by higher lightness/yellowness (L^*/b^*) ratio, sensory, and instrumental texture measurements along with oscillatory measurements (Fernández et al. 2006). The thickening effect observed in the study was attributed to the activity of PME.

The mathematical modeling of textural changes in potatoes during blanching is important in order to design blanching processes. Liu and Scanlon (2007) investigated the influence of blanching time and blanching temperature on the texture of potato strips and built a model to serve as a guide to potato blanching operators (Eq. 2.1):

$$F_{\text{max}} = F_0 \exp[kt(1/T - 1/T_r)] \tag{2.1}$$

where F_{max} is the maximum force attained in the force–deformation curve (N), F_0 is the initial texture of strips (N), k is a constant, t is the blanching time (min), T is the blanching temperature (°C), and T_r is a reference temperature at which no textural change with blanching time would be expected to occur (°C). Such model was validated experimentally suggesting that it can be used for estimating potato texture on the basis of blanching time and blanching temperature.

Moyano et al. (2007) used a model based on two irreversible serial chemical reactions for fitting the effect of blanching time on the texture of potato cubes. Equation 2.2 shows the two irreversible chemical reactions that are believed to occur during potato cubes blanching:

$$R(\text{raw tissue}) \xrightarrow{k_1} S(\text{soft tissue}) \xrightarrow{k_2} H(\text{hard tissue}) \tag{2.2}$$

where k_1 and k_2 represent the specific disappearance rate of the raw tissue and the specific disappearance rate of the soft tissue, respectively. After assuming that the concentration of these three types of tissue will change during blanching according to first-order kinetics and performing some calculation steps, the following model relating texture to blanching time was proposed:

$$F_{\text{MAX}}^* = \frac{F_{\text{MAX}}}{F_{\text{MAX0}}} = 1 - K[1 - \exp(-k_1 t)] \tag{2.3}$$

where F_{MAX}^* is the dimensionless maximum force; F_{MAX} is the same parameter defined as F_{max} in Eq. 2.1; t is the blanching time (s); F_{MAX0} is the maximum force at $t = 0$; K is a proportional constant linking potato texture with dimensionless concentration of soft tissue; and k_1 is the same k_1 as in Eq. 2.2. This model was superior to the traditional first-order kinetic model for explaining the changes in potato texture during blanching between 50 and 100 °C.

Color and oil uptake are key-quality parameters of fried potatoes. Studies dealing with the measurement of these parameters as affected by blanching are numerous. For instance, potato strips blanched at 97 °C for 2 min and subsequently fried presented impaired color, loss of textural quality, and increased oil absorption as compared to unblanched ones (Alvarez et al. 2000). On the other hand, blanching at 75 °C for 8 min followed by immersion in sucrose–NaCl aqueous solutions reduced oil uptake by 15% in potato strips during ulterior frying (Moyano and Berna 2002). They explained such effect by the fact that, during frying, solutes are concentrated on the surface, enhancing the formation of a crust that acts as barrier to oil uptake. In another study, blanching in 0.5% $CaCl_2$ solution at 85 °C for 6 min followed by immersion in 1% aqueous solution of carboxymethyl cellulose type hydrocolloids reduced oil uptake in fried potato strips in 54% (Rimac-Brnčić et al. 2004). This effect was attributed to the formation of a fine net structure by hydrocolloids and calcium chloride which prevents oil absorption by the potato strips during frying. Another report showed that blanching of potato chips at 85 °C for 3 min followed by air drying at 60 °C at 1 m/s until they reached a moisture content of 60% (w.b.) reduced oil absorption by 20% during subsequent frying, while blanching alone increased oil absorption (Pedreschi et al. 2005). This behavior could be related to gelatinization of starch during blanching (Califano and Calvelo 1987) along with increased dry matter content in the chips after drying, which is related to lower oil uptake (Zorzella et al. 2003). Blanching in water at 85 °C for 5 min followed by soaking in 20 g/L NaCl solutions at 25 °C for 5 min was more effective in reducing oil uptake in potato chips than blanching followed by coating with a 16 g/L edible film (hydroxypropyl methylcellulose) solution (Durán et al. 2007). Digital image analyses revealed that blanching at 70 °C for 5 min followed by soaking in NaCl solutions (0.6–9.0%) produced paler potato chips (Fig. 2.3) (Santis et al. 2007). Additionally, blanching at 85 °C for 3 min followed by soaking in 3% NaCl solution at 25 °C for 3 min reduced oil absorption during frying and produced paler and crispier chips when compared to the use of blanching alone (Pedreschi et al. 2007).

Another study on this issue showed that decrease in oil uptake by potato sticks was achieved by blanching at 50 or 85 °C for 5 min in 0.1% ascorbic acid solutions, while preserved color was observed after blanching above 69 °C and reduced product water activity was obtained for all blanching conditions (Reis et al. 2008). The effect on oil uptake can be attributed to the gelatinization of starch and to the activation of PME which leads to decrease in porosity. The preservative effect on color was due to the inactivation of PPO. The effect in water activity might be explained by freer moisture removal during frying resultant from loss of osmotic integrity and cellular damage produced during blanching (Alvarez et al. 2000). Low-temperature long-time blanching (60 °C/40 min) was effective in producing potato chips with less absorbed oil and of preserved color (Pedreschi et al. 2009). This behavior was, respectively, attributed to the activation of PME and to leaching of reducing sugars and amino acids that inhibited non-enzymatic browning.

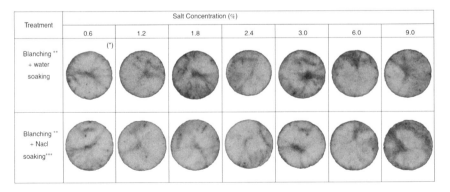

Fig. 2.3 Examples of images of potato chips previously blanched (70 °C for 5 min) and soaked at 25 °C for 5 min in solutions of different NaCl concentrations (Santis et al. 2007)

Dried potato products are also appealing to consumers due to their convenience of storage, transportation, and preparation. They can be easily rehydrated at home giving origin to potato pieces and purees. Blanching has been used before drying of potato products aiming at improving their final quality. For example, Cunningham et al. (2008) studied the texture changes during rehydration of dried potato cylinders. Their study showed that rehydration of dried potato cylinders was improved by a previous blanching treatment, ultimately resulting in a softer texture. They attributed this behavior to a complete starch gelatinization obtained by blanching at 100 °C for 5 min. Carillo et al. (2009) found that blanching in water at 100 °C for 3 min followed by drum drying resulted in potato powder of preserved color and preserved levels of starch, total proteins, and free amino acids. In addition, they found that the reconstituted product showed consistency similar to the fresh puree. The preserved color was probably due to the inactivation of the enzymes responsible for enzymatic browning during blanching, while the proper consistency might be related to the high level of starch, a thickening agent, in the reconstituted puree.

Another class of potato products under study is the class of the ready-to-cook and ready-to-fry products. Such products are neither completely processed nor fresh. They combine convenience with extended shelf life based on operations such as peeling, cutting, blanching, and modified packaging. Montouto-Graña et al. (2011) observed that the use of chemical pretreatments with citric acid and ascorbic acid was preferable to blanching for maintaining the quality of vacuum-packed ready-to-cook potatoes. They concluded that blanching at 90–100 °C conferred boiled appearance to the potatoes, especially at longer blanching times (5 min). Oner et al. (2011) found that blanching in 0.5% $CaCl_2$ solution at 60 °C for 10 min and then in boiling water (~ 98 °C) for 5 min combined with in-package gaseous ozone treatment increased the shelf life of potato strips during refrigerated storage. They observed that blanching was crucial for color preservation, which was attributed to inactivation of phenolase enzyme claimed to be responsible for enzymatic browning in potatoes. Oner and Walker (2011) also tested the effect of blanching combined with a technology called near-aseptic packaging on the shelf life of refrigerated potato strips. They found that blanching was effective in reducing microbial load to undetectable levels, and providing proper color and texture to the product after frying, especially when the blanching conditions were 20 min at 60 °C in 0.5 $CaCl_2$ solution followed by 5 min in boiling water.

Summarizing, potatoes are usually blanched for improving sensory characteristics like color and texture and also to reduce oil uptake in fried products. The inactivations of PPO and possibly POD along with the activation of PME are claimed to be behind color preservation and texture improvement. Decrease in oil uptake in fried potatoes due to blanching has also been ascribed to the activation of PME and also to the gelatinization of starch. These important goals for potato processors have been increasingly achieved by means of optimized blanching processes. Nevertheless, potatoes are not the only foods subjected to blanching. Several other foods have been blanched in an attempt to improve quality. The next section is devoted to this issue.

Impact of Blanching on the Quality of Miscellaneous Foods

In this section, the blanching processes of miscellaneous foods are dealt with as based on articles published over the last 10 years. More specifically, the works are presented chronologically and alphabetically but when two or more articles on a specific food were found, they were presented together aiming at showing the advances on the blanching of this specific food. Even though blanching is usually applied to vegetables and vegetable pieces, reports on the blanching of seafood, fungi, vegetable pastes, and fruits are available. They will be presented below.

Afoakwa et al. (2006) found that blanching cowpeas in boiling water for 5 min preceded by soaking in water containing 0.5% of sodium hexametaphosphate for 12 h improved the quality of canned cowpeas. A proper quality of canned cowpeas was represented by low moisture content, low pH, reduced leaching of solids, high drained weight, low number of split seeds, and intermediate hardness. Low moisture content and low pH inhibit microbial development thus increasing product shelf life, while reduced loss of solids reflects proper nutritional quality, and high drained weight, low seeds splitting, and hardness reflect proper sensory quality. The processing conditions that led to canned cowpeas of good quality were at intermediate levels of the experimental design.

Blanching pretreatment in saline solution improved the quality of two varieties of truffles according to a study by Al-Ruqaie (2006). In his study, two NaCl concentrations, 2 or 4%, and two blanching times, 2 or 4 min, were tested. Vinegar spraying was tested as well. The truffles blanched in 4% boiling NaCl solution for 4 min received higher sensory scores for color, texture, and flavor when compared to the unblanched truffles. The product was either frozen or dried after blanching. In this sense, blanching was crucial to the preservation of the product quality after 1 year of storage since neither the frozen nor the dried truffles presented a minimally acceptable quality when they were not blanched. Such effect of blanching on quality maintenance was probably due to enzymatic inactivation and microbial population reduction ultimately leading to reduced deterioration of sensory quality during storage.

Desalted cod is a chilled product presenting a shelf life of a few days. Fernández-Segovia et al. (2006) proposed the use of blanching or additives combined with vacuum or modified atmosphere packaging to increase desalted cod shelf life. They found that the most representative volatile for expressing desalted cod shelf life was 3-methyl-1-butanol since it is associated with unpleasant spoiled fish odor. Blanching presented no effect on the volatile composition of desalted cod immediately after treatment. On the other hand, after 42 days of storage at 4 °C the concentration of 3-methyl-1-butanol was lower in blanched cod than in control. Furthermore, the concentration of ketones and aldehydes described as components of the aroma of fresh fish was higher in blanched samples than in control. Despite the fact that the use of additives combined with modified atmosphere packaging was considered the best strategy for increasing desalted cod shelf life, blanching in boiling water for 1 min combined with vacuum or modified atmosphere packaging was also considered to be effective for that purpose. The positive effect of blanching on the odor of desalted cod could be attributed to the inhibitory effect of heating on microorganisms responsible for spoilage in fish. Another study showed that blanching of desalted cod in boiling water for 1 min combined with vacuum packaging increased product shelf life as expressed in terms of microbiological counts, total volatile basic nitrogen (TVB-N) concentration, and sensory scores (Fernández-Segovia et al. 2007). The effect of blanching on microbiological counts was due to the effect of heat on microorganisms, which in fact is related to the maintenance of TVB-N at acceptable levels, since TVB-N contents increase with the beginning of microbial spoilage (De Jesus et al. 2001).

Ismail and Revathi (2006) tested several hot water blanching conditions to find out which of them was able to inactivate POD and lipoxygenase (LOX) in chili puree prior to canning. Both enzymes were considered as unfavorable for the quality of the product due to the fact that they produce off-flavor and release intracellular gases of the plant tissue. They found that blanching at 100 °C for 1 min was enough to inactivate LOX, while the effective inactivation of POD required 6 min at 100 °C (Table 2.1).

Since the goal was to inactivate both enzymes, the latter conditions were recommended for blanching chili puree prior to canning. According to Whitaker (1996), even though most enzymologists do not consider elevated temperature to be an enzyme inhibitor, heating can decrease enzyme activity by denaturing some of the enzyme.

Other reports on the blanching of chili showed that blanching prior to spice manufacturing yielded a product with a moderate pungency while reducing the microbial load and preserving color (Schweiggert et al. 2005, 2006). They tried two methods of processing: blanching the chilies before mincing or mincing the chilies and then heating the paste. Blanching and heating conditions were the same, viz.,

Table 2.1 Reaction rate constant for peroxidase activity of *Capsicum annum var Kulai* blanched at various temperatures (Ismail and Revathi 2006)

Blanching time, min	Reaction rate constant/g sample for peroxidase activity		
	90 °C	95 °C	100 °C
0 (unblanched sample)	0.205 ± 0.12	0.205 ± 0.25	0.205 ± 0.23
1 min	0.195 ± 0.01	0.190 ± 0.24	0.183 ± 0.12
2 min	0.188 ± 0.56	0.182 ± 0.36	0.174 ± 0.01
3 min	0.176 ± 0.78	0.171 ± 0.01	0.15 ± 0.23
4 min	0.165 ± 0.15	0.159 ± 0.54	0.08 ± 0.56
5 min	0.144 ± 0.54	0.149 ± 0.12	0.025 ± 0.88
6 min	0.114 ± 0.36	0.095 ± 0.65	0.010 ± 0.01
7 min	0.085 ± 0.63	0.065 ± 0.14	0.010 ± 0.20
8 min	0.056 ± 0.56	0.051 ± 0.12	ND
9 min	0.032 ± 0.25	0.014 ± 0.03	ND
10 min	0.012 ± 0.45	ND	ND

Mean ± standard deviation based on three measurements; *ND* not detected

80, 90, or 100 °C during 5 or 10 min. Blanching before mincing was found to degrade less capsaicinoids, which are the compounds responsible for the peppers pungency. A blanching temperature of 90 or 100 °C and a blanching time of 5 min were especially appropriate in this sense. Hot chili pepper POD remained active even after blanching under severe conditions (100 °C/10 min). The moderate pungency was probably achieved due to diffusion/leaching and thermal degradation of capsaicinoids during blanching. Decrease in the microbial load was due to the well-known lethal effect of heat on microorganisms. Finally, preserved color can be attributed to the inactivation of deteriorative enzymes in the tissue by the heat of blanching.

A report on the negative impact of blanching on a food nutritional value was published by Lisiewska et al. (2006), who found that blanching at 96–98 °C for 60–120 s reduced ash (20–25%), potassium (21–36%), and phytic phosphorus (16–38%) in grass pea seeds before canning or freezing. A minor reduction in magnesium and iron contents was also observed. Such effect could be explained by the leaching of soluble solids during hot water blanching.

Blanching of pearl millet grains at 98 °C for 30 s was found to be a suitable treatment for obtaining biscuits with proper nutritional and sensory quality (Singh et al. 2006). The biscuits prepared from pearl milled flour presented high contents of calcium, iron, manganese, and phosphorus along with low anti-nutrient content, high in vitro digestibility, and high sensory acceptability. In this case, blanching was important for leaching the anti-nutrient compounds while maintaining the nutritional and the sensory quality.

Important losses in the nutrient content of white cabbage after blanching were observed by Wennberg et al. (2006). The blanching treatment consisted in immersing shredded cabbage in boiling water for 5 min as part of the procedures of coleslaw preparation. According to the authors, soluble/insoluble fibers, low-molecular-weight carbohydrates, and glucosinolates were lost either by depolymerization, leaching into the blanching water or evaporation into the air. The fact that the cabbage was shredded facilitated heat transfer and thus enhanced the effect of blanching on the product.

Bambara groundnuts are an important source of protein in Africa. Blanching of Bambara groundnuts before canning was investigated in terms of the impact of blanching time on product nutritional and sensory quality (Afoakwa et al. 2007). Blanching was preceded by soaking in sodium hexametaphosphate and performed with boiling water during up to 12 min. Ash, solids, phytates, and tannin contents decreased with increasing blanching times, while moisture content increased slightly and hardness was not affected. Given the low leaching of nutrients and the high decrease in anti-nutrient compounds, viz., phytates and tannin, blanching was indicated as a pretreatment during canning of Bambara groundnuts. In fact, optimum pretreatment conditions were soaking in a 0.5% sodium hexametaphosphate for 12 h followed by blanching for 5 min.

Blanching of animal foods is not usual. In one of the few reports on this issue, Sreenath et al. (2008) found that blanching of shrimps prior to canning led to a tougher texture in the product. A texture profile analysis (Fig. 2.4a) and a shear

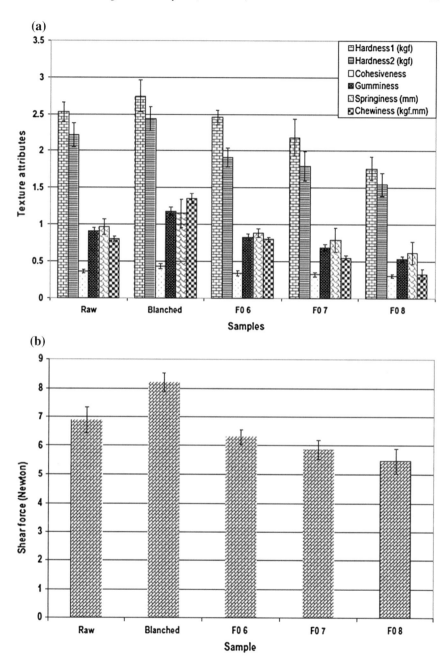

Fig. 2.4 **a** Hardness-1 (first compression), Hardness-2 (second compression), cohesiveness, springiness, gumminess and chewiness of raw, blanched, and processed (retorted at 121.1 °C for 6, 7, or 8 min) shrimp muscle (Sreenath et al. 2008). **b** Shear strength of raw, blanched, and processed (retorted at 121.1 °C for 6, 7, or 8 min) shrimp muscle. Results are mean ± SD, $n = 7$ (Sreenath et al. 2008)

force measurement (Fig. 2.4b) were used to support the conclusion. Blanching was performed in water heated at 80 °C containing 3% of sodium chloride for about 7 min. The firming effect of blanching on shrimp meat was attributed to thermal shrinkage or muscle denaturation and moisture loss. Nevertheless, subsequent sterilization promoted softening of shrimp meat and, under optimized sterilization conditions, canned shrimp with proper sensory quality curry could be obtained.

The kinetics of POD inactivation during blanching of pumpkin obeyed a first-order model (Eq. 2.4), while color and texture changes followed a fractional conversion kinetics model (Eq. 2.5), both respectively shown below (Gonçalves et al. 2007):

$$\frac{P}{P_0} = \exp\left(-k_{(T)}t\right) \tag{2.4}$$

$$\frac{P - P_{eq}}{P_0 - P_{eq}} = \exp\left(-k_{(T)}t\right) \tag{2.5}$$

where P is any measured quality factor, e.g., POD activity, a color parameter or a texture parameter; the index 0 indicates the initial value; the subscript eq indicates equilibrium value; t is the blanching time; and k is the rate constant at temperature T. Blanching temperature ranged between 75 and 95 °C. Blanching at 90 °C for 5.8 min and blanching at 95 °C for 3.9 min were able to satisfactorily reduce the POD activity, viz., 90% of reduction, while preserving pumpkin color, viz., L^*, redness (a^*), b^*, and chroma (C^*). Although the textural parameters firmness and energy were severely affected by blanching, the blanched pumpkin would be frozen and ultimately used for the manufacturing of soups or desserts, which made texture degradation irrelevant in this case. Dutta et al. (2009) observed that blanching in hot water increased the extractability/availability of carotenoids and gave a firmer texture to pumpkin. They attributed the increased extractability to increased tissue breakdown during blanching which might have disrupted carotenoid–protein complexes. On the other hand, the firming effect of blanching was probably due to the activation of PME in the tissue as discussed before in this chapter. They found that the best blanching conditions for obtaining an appropriate texture and increased carotenoid availability were a temperature of 55 °C for 3 min followed by storage at −18 °C. Gonçalves et al. (2011a) studied the effect of blanching before freezing of pumpkin, finding that blanching at 95 °C for 3.9 min significantly reduced firmness, L^*, a^*, b^*, vitamin C content, and POD activity. In addition, they observed that blanching altered the microstructure of pumpkin, leading to torn and irregularly shaped cells when compared to raw samples (Fig. 2.5). Possible mechanisms for the reported effects of blanching on the evaluated quality parameters include heat-induced cellular structure collapse, for the softening effect; non-enzymatic browning and carotenoid degradation, for color changes; leaching, for losses of vitamin C; and heat-induced denaturation, for the decrease in POD activity.

Gowen et al. (2007), when studying the hydration of soybeans, found that a previous blanching in hot water at 100 °C for 1.5 min decreased the soybeans

(a)   (b)

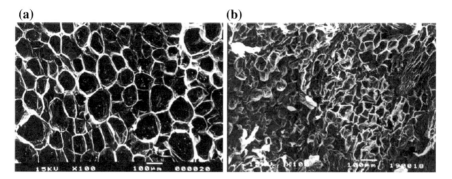

Fig. 2.5 Scanning electron micrographs of pumpkin cross section tissue at magnification 100×
and observation at 15 kV: **a** raw sample; **b** blanched sample at 95 °C for 3.9 min (Gonçalves et al.
2011a, modified)

soaking time by up to two hours, provided that soaking was performed in water
below 55 °C. They state that during the blanching process, the soybean seed coat is
plasticized. Plasticization by water increases molecular mobility (Roos 2008) and
reduces mechanical resistance in cellular foods (Corriadini and Peleg 2008),
allowing for faster moisture intake during the subsequent soaking process.
Furthermore, blanching reduced the soybeans microbial counts, which is an
important aspect to consider when hydrating grains at ambient temperature. Lv
et al. (2011) found that blanching of soybeans in water at 80–100 °C associated
with grinding reduced the concentration of undesirable beany flavors, being this
effect more pronounced for higher blanching times. Nevertheless, their results
suggest that the concentration of non-beany, desirable flavors also decreased with
higher blanching times. They recommended that an ideal blanching time, between 2
and 10 min, should be selected in order to eliminate as much beany flavors as
possible while maintaining a high level of non-beany flavors. The effect of
blanching on the decrease of aroma compounds of soybean might be attributed to
the volatilization of these compounds during heat treatment. Murugkar (2014)
investigated the effect of various blanching temperatures on the quality of soymilk
and tofu, finding that 100 °C was the optimal condition for grinding cum blanching
of sprouted soybeans. They report that products of preserved color, proper
viscosity/texture, high nutritional quality, and high sensory scores were obtained
when using the optimal blanching temperature. The mechanisms behind the
observed behavior may be, respectively, the reduced Maillard reaction, the gelation
of proteins, the optimal balance between anti-nutrients (trypsin inhibitor) and
nutrients content, and the proper color and texture obtained in the products. Peng
and Guo (2014) studied the effect of blanching of soybeans on the quality of
soymilk gels, finding that blanching at temperatures above 70 °C resulted in soymilk gels with desirable yogurt-like features. Their results were based on instrumental texture, sensory, calorimetry, and microscopy analyses. In this sense, they
found that high-temperature blanching reduced the textural parameters deformation

work and hardness, and the sensory parameters beany flavor, chalky taste, hardness, firmness, and mouth coating. In addition, this report showed that the microstructure changed from a "string of beads" to a "clustered protein aggregates" assembly after blanching above 70 °C (Fig. 2.6), which was possibly due to the denaturation of soy proteins.

Green peas are consumed in many countries where they are appreciated for their green color. Chlorophylls are usually the pigments responsible for green vegetables color. The degradation kinetics of chlorophylls *a* and *b* by heat and pH using green peas as model was studied by Koca et al. (2007). Heat treatment consisted of blanching at 70–100 °C in buffered aqueous solutions presenting pH of 5.5–7.5. Increase in blanching temperature and decrease in blanching solution pH led to increased degradation of both chlorophyll *a* and chlorophyll *b* as measured by high-performance liquid chromatography and instrumental color measurements. The temperature dependence of chlorophyll degradation and of color loss was well described by the Arrhenius equation, as follows:

$$k = k_0 \cdot \exp(-E_a/RT) \tag{2.6}$$

where, for a first-order kinetics, k is the rate constant (min^{-1}); k_0 is the pre-exponential factor (min^{-1}); E_a is the activation energy ($kJ\ mol^{-1}$); R is the universal gas constant ($8.315 \times 10^{-3}\ kJ\ mol^{-1}\ K^{-1}$); and T is the absolute temperature (K). They concluded that the effect of blanching temperature and pH on chlorophyll degradation was due to the increase in conversion of chlorophylls into pheophytins, changing the green color of vegetables to olive green during blanching at higher temperatures and lower pH. Gökmen et al. (2005) studied the effect of blanching on POD, LOX, vitamin C, chlorophylls, and color of peas during frozen storage. They found that blanching in water at 80 °C for 2 min was the best condition for promoting significant inactivation of the most resistant enzyme, i.e., 90% of POD inactivation, while maintaining high levels of vitamin C and chlorophylls, and preserving the original color. The positive effect of blanching on enzyme inactivation was due to heat-induced denaturation, while the loss of ascorbic acid may be attributed to leaching and the chlorophylls degradation may be attributed to their conversion to pheophytin, as described previously in this paragraph. Burande et al. (2008) studied the influence of pretreatments and fluidized bed drying conditions quality parameters of green peas. They found that blanching in a solution of 2.5% NaCl and 0.1% $NaHCO_3$ at 96 °C for 2 min combined with optimized drying conditions resulted in the highest rehydration ratio along with desirable texture, color, and appearance. The good rehydration of blanched foods may be attributed to heat-induced softening that facilitates the penetration of the rehydration water. This effect also produced desirable texture.

Kotwaliwale et al. (2007) reported undesirable effects of blanching in boiling water for 2 min on quality of dried oyster mushrooms, viz., higher darkening and higher hardening when compared to sulfite-treated samples. The textural changes were attributed to denaturation of protein during blanching along with filling of inter-molecular/capillary voids. The observed darkening could be related to the

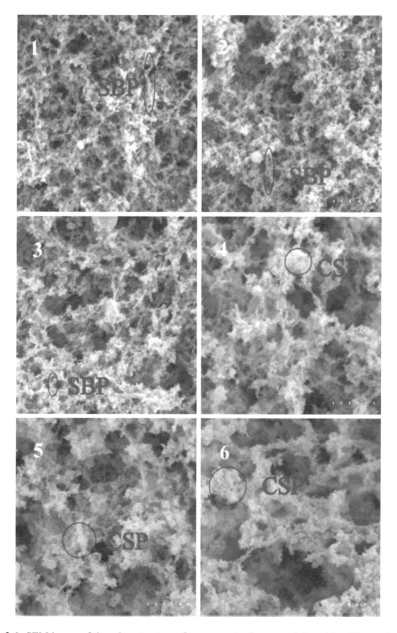

Fig. 2.6 SEM image of the microstructure of soy yogurt gels prepared from blanching soybeans under different temperatures. *1* CN-20, traditional soymilk preparation method. *2–6* Blanching at 50, 60, 70, 80, and 90 °C, respectively. Bars represent 10 μm. Networks of protein particles are in *white* and interspaces are in *black*. The different types of protein polymer in soy yogurt gels are highlighted with *red circles*. *SBP* string of bead protein polymer. *CSP* clustered stacked protein polymer (Peng and Guo 2014)

incomplete inactivation of enzymes responsible for enzymatic browning during blanching. Eissa et al. (2009) showed that blanching with saturated steam or boiling water before smoking was an excellent treatment for inhibiting enzymes and maintaining the microbiological quality of mushrooms. Their results suggest that both steam and water blanching for 7 min were able to inactivate PPO and POD and drastically reduce cellulase (CEL) activity. In addition, blanching for 3 min with water or vapor maintained the total bacterial and the yeasts and molds counts at acceptable levels after 8 weeks of refrigerated storage. Nevertheless, their study indicates that steam blanching was superior to hot water blanching with regard to color as measured instrumentally and by a sensory panel. The negative impact on color was attributed to heat-driven non-enzymatic browning which was more pronounced during hot water blanching than during steam blanching. Additionally, a negative impact of water blanching on taste, texture, odor, and appearance was observed, being related to protein breakdown and dehydration at the mushroom surface. Another study on mushrooms, conducted by Lespinard et al. (2009), led to the conclusion that blanching at 60–90 °C completely inactivated PPO in the tissue, while blanching at 50 °C only reduced its activity. Active PPO was claimed to be responsible for the high browning in the samples treated at 50 °C. In fact, they found that higher temperatures and lower times of blanching led to better color preservation, i.e., higher L^* value. They used an 18% shrinkage level to establish the blanching time, which was fixed at 77, 37, 23, 12, and 7 min, for blanching temperatures of 50, 60, 70, 80, and 90 °C, respectively. Their study also showed that blanching led to decrease in firmness irrespective of the temperatures used, which might be related to the heat-induced hydrolysis of the polymers which form the mycelium. Jaworska et al. (2010) investigated the effect of different blanching conditions on texture of the canned mushrooms *Agaricus bisporus* and *Boletus edulis*, finding that mushrooms pretreated by immersion in aqueous solutions at 96–98 °C containing 1% lactic acid and 0.1% L-ascorbic acid presented a tougher texture, i.e., higher values of force and work, when compared to those pretreated using pure water or aqueous solution of 0.5% citric acid and 0.1% L-ascorbic acid. Jaworska ct al. (2011) studied how blanching in water and in aqueous solutions of various substances and soaking-and-blanching in these solutions affected the amino acid content of frozen and canned mushrooms (*Pleurotus ostreatus*). They concluded that blanching at 96–98 °C for 3 min was superior to soaking for 1 h followed by blanching for the preservation of several mushroom amino acids in both the frozen and the canned product. Bernaś and Jaworska (2016) found that blanching in sodium metabisulfite (0.2%) and citric acid (0.5%) solution for 3 min at 96–98 °C was good for dry weight, vitamin B_3, and L-ascorbic acid retention in mushrooms.

Garlics can become pungent and greenish when not subjected to blanching during the preparation of pickled garlics, which stimulated Rejano et al. (2007) to study the blanching of garlics. Blanching was performed at 60–95 °C for times varying from 3 to 180 min aiming at pungency elimination and texture maintenance. As expected, texture was increasingly degraded with an increase in blanching temperature and time, which was attributed to loss of mechanical strength and to cell wall adhesion. At the same time, temperatures below 65.2 °C were not sufficient to eliminate pungency. Therefore, they found that blanching at 80–90 °C was the best strategy for eliminating pungency with minimal texture degradation in garlics. They attributed the effect of high-temperature blanching on pungency to thermal inactivation of enzymes responsible for producing pungent compounds. Another report on the blanching of garlic is provided by Fante and Noreña (2012), who studied the PPO, POD, and inulinase inactivation kinetics and the color changes in blanched garlic slices. The inactivation kinetics was well fitted by a biphasic first-order model (Ling and Lund 1978), as follows:

$$y = a_L \exp^{-k_L t} + b_R \exp^{-k_R t} \tag{2.7}$$

where y is the residual enzyme activity; k_L and k_R are the velocity constants of the heat-labile and heat-resistant components, respectively; a_L and b_R are the initial concentrations of the labile and resistant components, respectively; and t is the blanching time. They concluded that steam blanching at 100 °C was the best treatment for inactivating the selected enzymes while retaining high levels of inulin, glucose, and sucrose, and improving color. The heat of the steam was effective for denaturing enzymes thus inhibiting enzymatic browning and thereby preserving color. At the same time, steam blanching preserved sugars since it does not leach sugars as severely as hot water does.

There are cases where traditional methods like blanching are used in combination with emerging technologies with the aim to improve process efficiency and product quality. This is the case of the study carried out by Sarang et al. (2007) who tested the use of blanching to enhance the electrical conductivity of a solid–liquid food mixture and in this way improve the Ohmic heating of this system. The blanching solution consisted of a highly conductive sauce heated to 100 °C. This sauce was used for blanching various ingredients of a dish called "chicken chow mein" for variable times. In fact, the blanching time was optimized for all ingredients with the goal of maximizing their electrical conductivity, resulting in the following conditions: bean sprouts: 10 s; celery: 2 min; chestnut: 1.5 min; mushroom: 6 min; and chicken: 8 min. They concluded that blanching is suitable for improving the ohmic heating of this food system while maintaining its sensory quality. Wang and Sastry (1997) affirm that the cell wall breakdown and the changes in membranes and in other structures during heating treatments cause more mobile moisture, enhancing ionic mobility, and ultimately promoting increase in electrical conductivity.

Blanching was successfully used as a pretreatment for improving the quality of vanilla beans (Sreedhar et al. 2007). Hot water blanching at 63 °C for 3 min followed by scarification with a metal brush, immersion in 5 mg/L plant growth regulator (NAA) solution at room temperature (28–30 °C) for 5 min, and incubation at 38 °C for 10 days was found to be the best combination for curing the vanilla beans. Such processes led to satisfactory generation of flavoring compounds and proper texture in the beans while being faster than the conventional curing process. Blanching and the other pretreatments were claimed to enhance the vanilla flavor formation via regulation of β-glucosidase and CEL enzymes. Another report on the blanching of vanilla beans was provided by Van Dyk et al. (2010), who studied the effect of blanching on product quality features. They performed blanching at 67 °C for 3 min in hot water and subsequently they used specific curing treatments, finding that blanching produced beans of better appearance, namely a shiny brown-black color, when compared to unblanched beans. On the other hand, their results suggest that blanching inactivated the β-glucosidase enzyme responsible for converting glucovanillin into vanillin. Nevertheless, their blanched samples presented higher concentration of aromatic odors, which was attributed to the activity of other enzymes like POD. They suggested further studies to find mild blanching conditions able to disrupt the tissue in order to promote the mixing of glucovanillin and glucosidases while maintaining specific enzymatic activity for aroma generation in the vanilla beans. Once again, the positive effect of blanching on product color was confirmed along with the ability of the blanching heat to denature enzymes.

Fried sweet potato chips with proper textural quality were tentatively obtained by Taiwo and Baik (2007) using freezing, air drying, osmotic dehydration, and hot water blanching at 70 °C for 10 min as pretreatments. A positive effect of blanching on fried product cohesiveness, i.e., resistance to breakage under compression (Fig. 2.7a), and adhesiveness was observed (Fig. 2.7b). In this sense, blanched chips were more cohesive and less sticky. On the other hand, unblanched samples were less shrunk and puffier. It is not clear why blanching promoted such changes in the texture of sweet potato chips. A study on fried snacks showed that blanching of sweet potato slices and cubes in 1% sodium tripolyphosphate solution for 2 min at 100 °C produces restructured sweet potato sticks with improved quality (Utomo et al. 2008). They found that the aforementioned blanching conditions decreased fat absorption during frying, increased the ash content, and increased L^* of the final product as compared to hot water blanching. The mechanisms behind the effect of blanching on reduction of oil absorption and color preservation were discussed before in this paragraph for potatoes, while the increased ash content is related to the incorporation of sodium tripolyphosphate into the product during blanching. Conversely, blanching of orange-fleshed-sweet potato chips before sun drying promoted undesirable quality changes (Bechoff et al. 2010). The authors reported that blanching in boiling water (96 °C) for 8 or 11 min depending on the sweet potato variety used rendered the chips sticky due to starch gelatinization. He et al. (2014) investigated the effect of low-temperature blanching (LTB) on the activity of PME, polygalacturonase (PG) and β-amylase, firmness, and dry matter content of sweet potato slices. Furthermore, they assessed the degree of methylation of cell wall material extracted from the samples, the influence of the extracted PME on the hardness of a calcium-containing pectin gel, and the amount of free starch in the flour made with the samples. They found that blanching at 60 °C for 30 min was the best strategy for achieving a desirable PME and β-amylase activation along with PG inactivation which led to increased firmness in the tissue. Additionally, they concluded that these blanching conditions resulted in reduced loss of dry matter, higher gel hardness, and lower free starch content. Such behavior could be

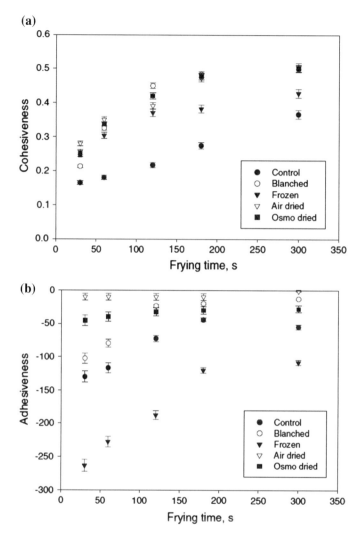

Fig. 2.7 Cohesiveness (**a**) and Adhesiveness (**b**) of sweet potato samples during deep fat frying at 170 °C (Taiwo and Baik 2007)

attributed to the activation of PME that increased slices firmness and gel hardness by means of mechanisms described previously in this chapter. The reduced loss of dry matter was due to the reduced blanching time when compared to the other blanching conditions used in the same work. Ndangui et al. (2014) found that blanching and CaCl₂ pretreatment differently affect the properties of sweet potato flour. Their results suggest that pretreatment of sweet potato slices with CaCl₂ at ambient temperature was the best strategy for obtaining high-quality sweet potato flour. Blanching was found to promote leaching of reducing sugars, which could impair the fermentation during the processing of bread with the obtained flour. Additionally, blanching was found to reduce $L*$, which could impair the processing of light-colored bread.

Coconut milk as extracted from blanched-frozen coconut kernel was found to present improved quality and extractability (Waisundara et al. 2007). Blanching was performed in hot water at 85–95 °C during 12–25 min. Blanched samples were subsequently sanitized with chlorine, vacuum packaged, and stored at −18 °C for up to eight weeks before the coconut milk extraction. Free fatty acids and peroxides, which reflect coconut milk degradation, showed very low values in all blanched samples. At the same time, the enzymes lipase and POD, which are associated with the generation of free fatty acids and peroxides, were inactivated by blanching. Figure 2.8 shows the effect of blanching and freezing on peroxidase activity of coconut kernels during storage. It was concluded that blanching was essential for producing coconut milk with a fresh appeal as compared to canned coconut milk.

According to Agüero et al. (2008), butternut squash is an important vegetable in the Argentinian cuisine, but its size and weight somehow impair its marketing. Therefore, they published a study on the minimal processing of this vegetable, taking into account that a blanching step is necessary to make butternut squash digestible. They found that several blanching conditions were able to inactivate 90% of the POD contained in the tissue, a value which is recommended for increasing shelf life while preserving most of the nutrients. Two forms of POD, namely the labile and the heat-resistant fractions, were active in the tissue. For blanching temperatures above 70 °C it was considered that the labile fraction of

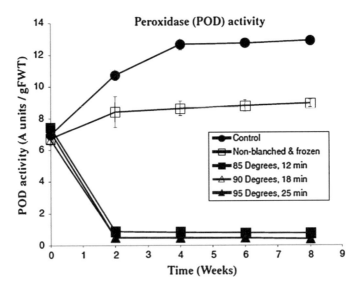

Fig. 2.8 Effect of different blanching treatments and frozen storage on the peroxidase activity of coconut kernels (Waisundara et al. 2007)

POD was quickly inactivated. Therefore, only the heat-resistant fraction was active. The inactivation kinetics for blanching temperatures above 70 °C followed a monophasic first-order model which when combined with the Arrhenius equation (Eq. 2.6) gives the following model:

$$RA = \exp\left\{-k_{ref}\exp\left[-\frac{E_a}{R}\left(\frac{1}{T} - \frac{1}{T_{ref}}\right)\right]t\right\} \tag{2.8}$$

where RA is the residual POD activity; k_{ref} is the inactivation rate constant (min^{-1}) at a reference temperature (T_{ref}); E_a is the activation energy for POD inactivation (J/mol); R is the universal gas constant (8.314 J/mol K); T is the temperature (K); t is the time (min); and T_{ref} is the mean of the tested blanching temperatures (358.15 K). For blanching temperatures below 70 °C, the two forms of POD were active, which makes the biphasic model more suitable for this case, as follows:

$$\begin{aligned} RA = f_L \exp\left\{-k_{L,ref}\exp\left[-\frac{E_{aL}}{R}\left(\frac{1}{T} - \frac{1}{T_{ref}}\right)\right]t\right\} \\ + f_R \exp\left\{-k_{R,ref}\exp\left[-\frac{E_{aR}}{R}\left(\frac{1}{T} - \frac{1}{T_{ref}}\right)\right]t\right\} \end{aligned} \tag{2.9}$$

where f_L and $k_{L,ref}$ (min^{-1}) are the initial fraction and the inactivation rate constant of the labile isoenzyme at the reference temperature, and f_R and $k_{R,ref}$ (min^{-1}) are those of the heat-resistant fraction; E_{aL} and E_{aR} are the activation energies for the inactivation of the heat-labile fraction and of the heat-resistant fraction (J/mol), respectively; and R, T, T_{ref}, and t are the same parameters presented in Eq. 2.8. In addition to the kinetic study, the impact of blanching on ascorbic acid retention was also assessed, leading to the conclusion that high-temperature short-time (HTST) blanching preserved more the original ascorbic acid of butternut squash when compared to longer processes at lower temperatures. In this case, the vitamin C may have been leached during blanching and the higher the blanching time, the more intense the leaching.

Cauliflower is a perishable food that is extensively consumed in several countries. Dehydration is a suitable technology for improving vegetables shelf life. In the study of Kadam et al. (2008), they investigated how blanching time, potassium meta bisulphite (KMS) concentration, and the use of different packaging materials affected the quality of dehydrated cauliflower during storage. The blanching and the KMS dipping were performed separately. They found that blanching of cauliflowers in boiling water during different process times did not affect their moisture content, while blanching for 3 min resulted in the best rehydration behavior and the best vitamin C retention. In addition, blanching for 9 min resulted in minimum non-enzymatic browning. These results might be attributed to the following mechanisms: a better rehydration behavior has been associated with a more preserved structure, which is possible when using short blanching times; the losses of vitamin C during blanching have been attributed to leaching, which is less intense for shorter blanching times; and the reduced non-enzymatic browning is related to increased leaching of reducing sugars and amino acids from cauliflower surface during long-time blanching.

The use of blanching–freezing–cooking for French beans (*Phaseolus vulgaris* L.) was not as proper as the use of cooking–freezing–microwave heating for preserving their amino acid content according to Kmiecik et al. (2008). In fact, they observed that blanching at 95–98 °C for 3 min followed by freezing and cooking resulted in total amino acids content 18.39% lower than the product prepared by cooking, freezing, and heating in microwave. As far as could be understood from the study, since the traditional method of preparation that involves blanching comprises two immersions in water, one possible explanation for these results is a higher leaching of amino acids in this method when compared to the cooking–freezing–microwave heating method that comprises only one immersion in water. Another work performed by the same group showed that this novel cooking–freezing–microwave heating method also preserves the mineral content of selected vegetables in a better way as compared to the blanching–freezing–cooking method (Lisiewska et al. 2008). They found that the content of both total ash and minerals such as P, K, Mg, Fe, Zn, and Mn was superior in French beans, broad beans, and peas treated by the novel method. Conversely, Słupski (2010) found that blanching–freezing–cooking was superior to cooking–freezing–microwave heating with concern to protein quality (essential amino acid index) and sensory quality of flageolet bean seeds (*Phaseolus vulgaris* L.). Nevertheless, he found that both methods of processing promoted a drastic decrease in tyrosine content, while the content of the other amino acids was differently affected for different cultivars.

Blanching of fruits is not usual, because most consumers prefer a fresh-like appearance and flavor in fruit products. Nevertheless, Kulkarni et al. (2008) studied the effect of blanching on the quality of dehydrated immature dates. They found that increasing blanching time between 5 and 20 min did not affect moisture, total sugars, and total soluble solids content, while vitamin C content decreased. Furthermore, their results showed that blanching for 15 min resulted in dehydrated dates of glossy appearance, soft texture, and high sensory acceptance. They concluded that blanching at 96 °C for 15 min improves the quality of dehydrated dates. The decrease in vitamin C may be attributed to leaching, while the softening effect is probably related to the effect of heat on the structural polysaccharides of the date fruits. In another study on fruit blanching, López-Malo and Palou (2008) used combined methods for preserving pineapple slices, observing that the use of blanching with saturated water vapor for 1 min was effective for reducing aerobic plate counts in about three-fourths in comparison to fresh pineapple. When combined with osmotic dehydration and immersion in preservative solution, the product was found to present microbiological and sensory stability during 3 months at 25 °C. Blanching did not impair color, texture, or sensory scores. The effect of blanching on microbial counts can be attributed to thermal death of microorganisms after vapor treatment. Other studies on the blanching of fruits will be presented in this chapter.

In mint leaves, the Weibull model was found to be the best one for describing the thermal inactivation of POD (Rudra Shalini et al. 2008). This model is presented below:

$$s = s_0\exp(-bt^n) \tag{2.10}$$

where s is the residual POD activity; s_0 is the initial residual POD activity; b is the scale factor (min^{-n}); n is the shape factor; and t is the heating time (s). The POD activity decreased as blanching proceeded, as expected. Blanching was carried out at 70–100 °C during 5 s–9 min. The appropriateness of the Weibull model was confirmed taking into account the low values of standard error and the high values of coefficient of determination (R^2). The temperature dependence of the parameter b which reflects the thermal reaction rate was well described by polynomial models. The inactivating effect of blanching on POD of several foods is well described and can be attributed to enzyme denaturation upon heating. This study contributed for the designing of a blanching process for mint leaves. It is interesting to mention that POD is seldom totally inactivated since this would cause over-blanching.

Blanching is also a way to improve the quality of fried yam slices, as shown by Sobukola et al. (2008). They found that blanching in water at 70–75 °C for 4–5 min reduced the oil uptake during the posterior frying. In addition, blanching produced proper crispness and color in the final product. The decrease in oil uptake may be attributed to mechanisms like starch gelatinization and PME activation, as discussed before in this chapter. The favorable effect of blanching on texture is possibly due to the softening effect of blanching thus reducing the breaking force. The ideal color of yams blanched at optimal conditions was obtained probably due to the partial removal of reducing sugars from yam slices surface thus controlling non-enzymatic browning. In another report, Abiodun and Akinoso (2015) blanched trifoliate yam dices at 65 °C for 10 min before the preparation of yam flour and yam dough. They found that the blanched flour presented higher swelling index and higher water holding capacity, suggesting that blanching improves the functional properties of trifoliate yam flour.

For hazelnut meals, blanching of hazelnut kernels was found to deaccelerate the generation of free fatty acids during storage, an effect that was attributed to partial enzymatic inactivation of lipase (Cam and Kilic 2009). Nevertheless, the authors observed that blanching seems to promote oxidation of hazelnut lipids, since the oil stability index was lower for the oil extracted from blanched samples at the beginning of storage. They hypothesized that this reduced oil stability index was due to the fact that the blanched hazelnuts water activity was below that of the water monolayer thus exposing their lipids directly to oxygen. After 10d of storage and until the end of the 50d storage, though, they found that the oil extracted from blanched samples presented higher oil stability index than the oil extracted from unblanched samples (Fig. 2.9).

Vegetable soybeans that were blanched for 2 min in a steam-jacketed kettle presented 80% of inactivation of trypsin inhibitors while maintaining proper color, texture, and sucrose levels (Mozzoni et al. 2009). The authors compared the use of

Fig. 2.9 Oil stability index of oils extracted from hazelnut meal samples during storage (Cam and Kilic 2009)

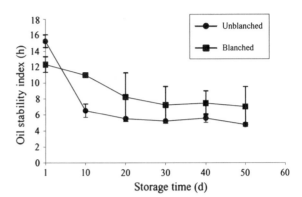

cooking in a household stove equipped with an electrical resistance and blanching in a steam-jacketed kettle, the former simulating home-cooking conditions and the latter simulating an industrial water blanching process. According to the authors, the inactivation of trypsin inhibitors was due to the effect of heat. In addition, they attributed the firming effect of blanching during the first minute of process to starch gelatinization and reaction between Ca^{2+} released from the cells and pectin. Finally, the short blanching process did not promote discoloration or leaching of sugars, which are common drawbacks of longer blanching processes.

A report on mango slices showed that steam blanching reduced the activity or completely inactivated PPO and POD depending on the blanching time used (Ndiaye et al. 2009). They found that such inhibiting effect reflected on color of mango puree made from blanched slices, which was preserved after 20 days of refrigerated storage. In fact, their results showed that blanching the 1-cm-thick slices with saturated vapor for 3, 5, or 7 min was the best treatment in this sense. On the other hand, they found that samples blanched for 1 min or unblanched samples added with 1% ascorbic acid suffered enzymatic browning during storage, which was correlated with the activity of PPO and POD. The enzymatic browning in mango was denoted by low values of L^*, high positive values of a^*, and low positive values of b^*. Liu et al. (2014) used a headspace fingerprinting technique to show that blanching significantly affected the volatile profile of mango nectar. In fact, they found that the concentration of most of the volatiles was lower in blanched nectar, which was attributed to thermal induced degradation, in the case of terpenoids, and to thermal inactivation of enzymes responsible for generating hexanal from unsaturated fatty acids, in the case of hexanal. They performed blanching at 95 °C for 8 min, a severe condition when compared to the conditions used for the subsequent treatments, viz., high pressure and pasteurization. Then, they concluded that blanching was the main responsible for the changes in the volatile fraction of mango nectar.

Carrots are an important food in several countries and their processing into the frozen, pureed, dried, and other forms may include a blanching step. The next four paragraphs will be dealing with carrots blanching.

Shivhare et al. (2009) studied the effect of various blanching temperatures, times, and solutes on catalase and POD enzymatic activity, β-carotene, and vitamin C contents along with juice yield. In addition, they modeled the POD inactivation. Their results showed that water blanching at 95 °C for 5 min was preferred to steam blanching or blanching in acetic acid solution or calcium chloride solution at 80–100 °C during 2–10 min with regard to the impact on the selected quality parameters. When using the former blanching conditions, they obtained at the same time complete enzymatic inactivation, high carotene (Fig. 2.10a) and vitamin C content (Fig. 2.10b), and high yield of carrot juice. The Weibull model (Eq. 2.10) was shown to be the most appropriate for expressing the changes in POD activity during blanching (Fig. 2.11). It can be inferred that the heat denatured POD after 5 min at 95 °C. The same blanching conditions were favorable for β-carotene and vitamin C retention and yield of carrot juice because this is a relatively short time which is not enough to promote severe losses of these compounds and of juice into the blanching water. Another study on carrots dealt with the kinetics of POD inactivation, phenolic compounds degradation, color, and texture changes during

Fig. 2.10 Effect of blanching on β-carotene (**a**) and vitamin C (**b**) content of carrots (Shivhare et al. 2009)

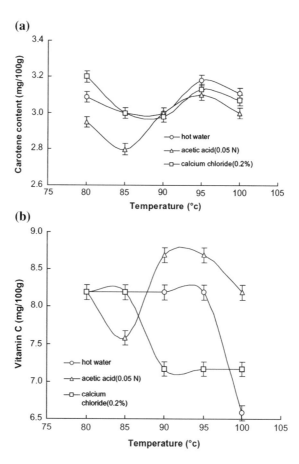

Fig. 2.11 The activity curve of peroxidase in carrot extract fitted with the Weibull distribution at 80–100 °C (Shivhare et al. 2009)

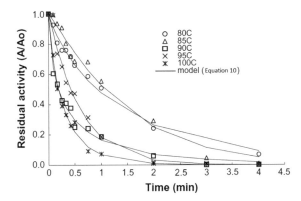

blanching (Gonçalves et al. 2010). They found that the changes in POD activity and phenolic content obeyed first-order kinetics (Eq. 2.4), while the changes in color and texture obeyed a fractional conversion model (Eq. 2.5). Additionally, they observed that the temperature dependence of the studied parameters was well described by the Arrhenius equation (Eq. 2.6). They came to the conclusion that blanching at 80 °C for 6 min promoted the desired 90% inactivation of POD while preserving phenolics, color, and texture.

Tansey et al. (2010) used two-step blanching during the preparation of *sous vide* frozen carrots, more specifically the immersion of carrot slices in water at 50 °C for 30 s followed by immersion in water at 90 °C for 3 min. They found that optimal product texture can be achieved by means of the proposed blanching treatments followed by cooking and freezing. Such achievement was attributed to the firming effect of LTB that activates the PME enzyme and the softening effect of HTST blanching, cooking, and freezing on carrot cell structure. At the same time, they observed that soluble solids were lost from the carrots during blanching due to leaching. Neri et al. (2011) studied the effect of blanching in water and in sugar solutions on microstructural and macrostructural characteristics of carrots, finding that maltose and trehalose exert a protective effect on carrot structure. They confirmed this protective effect by means of cryo-scanning electron microscopy techniques. Additionally, they found that blanching at 75 °C increased carrot hardness while blanching at 90 °C reduced hardness. The hypotheses contained in the report for explaining the observed behavior were: the protective effect of sugars on carrot structure during blanching was related to sugar uptake and decrease in solute leaching; the hardening effect of blanching on carrot at lower blanching temperature was due to the activation of PME in the tissue; and the softening effect of blanching on carrot at higher blanching temperature was due to dissolution of pectin, gelation of cell wall polymers and cells separation.

Lemmens et al. (2013) investigated the effect of blanching on texture and β-carotene bioaccessibility of carrots before thermal treatments (pasteurization or sterilization). They concluded that blanching at 60 °C for 40 min followed by immersion in Ca^{2+} solutions improved carrot texture by preventing severe hardness decrease after thermal treatments, being this preventive effect more pronounced at mild thermal treatment conditions. This behavior is probably due to the activation of PME in the tissue during LTB, as discussed previously. At the same time, though, blanching reduced β-carotene bioaccessibility, which was attributed to the fact that Ca^{2+} impairs the formation of micelles that are the forms in which β-carotene is absorbed by the body. A study on pasteurized carrot juice showed that acid blanching reduced and stabilized the pH and improved the cloud stability of the product (Yu and Rupasinghe 2013). The effect of acid blanching on the initial decrease of pH was due to the presence of 2% of citric acid in the blanching solution, while the pH stabilization during storage was due to the lethal effect of heat on spoilage microorganisms that would ferment sugars and decrease the pH. The influence of acid blanching on the cloud stability (absence of sedimentation) of carrot juice during storage was attributed to the inactivation of microorganisms and to the denaturation of PME which would cause sedimentation during storage.

Neri et al. (2014) studied the effect of blanching in water, trehalose, and maltose solutions on the texture, microstructure, POD, and pectinesterase (PE) activities of carrot slices before freezing. They concluded that LTB (75 °C) was not able to inactivate POD and PE even when using long blanching times (10 min), while high-temperature blanching (90 °C) inactivated POD and PE even when using short blanching times (3 min). With regard to texture, they found that blanching either in water or in sugar solutions reduced the maximum load (N) of the samples when compared to unblanched samples. Additionally, the carrot microstructure was negatively affected by long-time blanching, being this effect attenuated by the addition of sugars to the blanching solution. The inactivating effect of high temperature on enzymes is due to the denaturation. The softening effect of blanching was attributed to the residual activity of PE, in the case of LTB, and to the heat-induced cellular damage and loss of turgor in the case of high-temperature blanching. The protective effect of sugars on the carrot microstructure was ascribed to sugar penetration in the vegetable tissue. Ma et al. (2015) blanched carrots at 86 °C for 10 min before the extraction of their essential oil and evaluated the effects of this pretreatment on the oil chemical composition and antimicrobial activity. Their results clearly show that blanching decreases the essential oil components and promotes conversion of some compounds into others. Furthermore, they found that blanching weakened the antimicrobial effect of carrot essential oil. The decrease in the essential oil components is probably due to the volatilization of aroma compounds during carrot blanching. Such lost compounds may be responsible by the antimicrobial activity of the carrot essential oil, while the compounds generated during blanching may not present a relevant inhibitory effect against microorganisms. Paciulli et al. (2016) found that a blanching treatment prior to freezing and cooking avoided the enzymatic browning and stabilized β-carotene in carrots, which was attributed to the inactivation of oxidative enzymes.

In fish (*Labeo rohita*) eggs, blanching previously to pickle preparation was found to provide the product with good chewability and reduced fish odor while improving its shelf life (Balaswamy et al. 2010). They performed blanching at 110 °C for 5 min in saline solutions of different concentrations, concluding that the

1 M NaCl solution was the best treatment. Combined with the use of an acidic pH, blanching was responsible for the great shelf stability of the product as denoted by low microbial load (Total Plate Count) and high sensory scores after 6 months of storage at 28 °C. The effect of blanching on microbial load can be attributed to the lethal effect of heat on microorganisms ultimately leading to absence of off-flavors after 6 months of storage. The good chewability of the blanched product could be related to protein coagulation. The reduced fish odor in the blanched product could be attributed to the removal of aldehydes responsible for such odor by volatilization during blanching.

Apples were subjected to blanching or freezing prior to drying by different methods and their effect on dried product quality was evaluated (Acevedo et al. 2008). The results showed that blanching followed by vacuum drying gave origin to dried apples with sorption behavior similar to vacuum-dried/unblanched samples and to slow frozen/freeze-dried samples. With regard to microstructure, the authors found that both blanched and unblanched vacuum-dried samples presented high level of structural collapse and tissue disruption, being these changes more pronounced at high storage temperature. With regard to color, their results show that blanched samples were lighter than unblanched samples during storage at 70 °C, which can be attributed to the inhibition of enzymatic browning in blanched samples. On the other hand, blanching did not affect the texture of vacuum-dried samples. A comprehensive study on apples showed that the color of dried apples is affected by both enzymatic and non-enzymatic browning, being the rate of both affected by blanching (Lavelli and Caronni 2010). They observed that at a storage temperature of 20 °C and water activity (a_w) below 0.32, unblanched and blanched freeze-dried apples both presented low browning, while at a_w above 0.32 there was the occurrence of enzymatic browning in unblanched samples. On the other hand, they found that at a storage temperature of 40 °C blanching promoted the Maillard reaction, whose rate increased with increasing a_w. The absence of enzymatic browning in blanched samples was due to the denaturation of PPO caused by heat. The increase in the rate of Maillard reaction with temperature is well described in various food systems, while the relationship between loss of cellular integrity and browning was reported by Acevedo et al. (2008), which might help to explain the higher Maillard browning in blanched samples stored at 40 °C.

In asparagus, Słupski et al. (2010a) showed that blanching at 96–98 °C for 4 min prior to freezing and cooking led to higher losses of tyrosine and aspartic acid when compared to cooking–freezing–thawing in microwave oven. Nevertheless, the quality of the protein in asparagus treated by both processes was similar. For the remaining amino acids, decreases or increases were noted, with values ranging from −14% to +16% in the evaluated samples. The same fashion of study was conducted on New Zealand spinach (*Tetragonia tetragonioides* Pall. Kuntze) by Słupski et al. (2010b), who concluded that blanching increased the samples dry matter content, which indeed led to increase in the relative content of most of the amino acids in blanched samples.

Almonds are appreciated in several countries due to their pleasant sensory characteristics and nutritional value. Blanching of almonds was found to provide them with higher values of L^* when compared to oil- and air-roasted samples, which mean that the blanched almonds presented lighter color (Altan et al. 2011). In addition, they observed that blanched almonds presented a more preserved cellular structure as compared to oil-roasted almonds, which would lead to less release of oil during storage.

Blanching of broccoli was found to inactivate 90% of initial POD activity while maintaining a green color, i.e., a hue angle (h^*) of 137.95, and high levels of vitamin C (36.07 mg/100 g) (Gonçalves et al. 2011b). They found that the blanched-frozen broccoli presented a shelf life of about 4 months, which may be explained by enzyme denaturation and microbial death due to the heat (70 °C for 6.5 min) of the blanching water. Another study showed that broccoli purée presented different behaviors after blanching at low or high temperature, namely the occurrence or the absence of syneresis, respectively (Christiaens et al. 2012). Additionally, LTB (60 °C/40 min) increased the physical resistance to blending which was attributed to increased adhesion between cells, ultimately resulting in a higher quantity of larger particles in the low-temperature blanched purée. This behavior can be attributed to the activation of PME enzyme in the tissue, which was confirmed by the decrease in the degree of esterification of pectin of the low-temperature blanched broccoli in that study. On the other hand, high-temperature blanching (95 °C/5 min) promoted only a small decrease in the degree of esterification of pectin, which was attributed to chemical de-esterification and resulted in a stable purée, i.e., without separation between serum and pulp. Jin et al. (2014) studied the sorption isotherms of blanched and unblanched pre-dried broccoli. They found that blanching in water at 90 °C for 3 min leached 49.8% of the glucose from the florets and 28.5% of the glucose from the stalks. Additionally, blanching was said to denature proteins and break cell walls, which in turn promoted a higher release of bounded water when compared to unblanched samples.

Lisiewska et al. (2011) compared the effect of their innovative method for vegetable processing, tested before in other foods, for spinach preparation. As mentioned before in this chapter, the new method consists in cooking–freezing–refrigerated storage, followed by preparation for consumption in a microwave oven. Blanching–freezing–refrigerated storage followed by cooking was shown to be roughly similar their innovative method with regard to product amino acid content, with both methods generally increasing the content of selected amino acids as compared to the raw material. According to the authors, during heat treatment, the raw material can shrink, releasing water and, despite decreased levels of constituents, its dry matter content increases, which helps to explain the increase in amino acid concentration in the blanched and in the cooked product.

Abalone, a marine gastropod, was treated by blanching in hot water at 100 °C for 1–4 min before soaking in a typical Korean sauce and storing under refrigeration (Moon et al. 2011). The authors found that blanching for 1 min resulted in the highest sensory scores for taste, flavor, and texture while unblanched abalone presented strong fishy smell and hard texture, and abalone blanched for longer times presented overcooked smell and softer texture. The mechanism behind the described behavior is probably the leaching and volatilization of fishy smelling compounds, which impacted taste and flavor along with the heat-induced protein denaturation which impacted texture.

Prajapati et al. (2011) found that blanching of Indian gooseberry (aonla) shreds prior to drying was, in general terms, inferior to blanching in 0.1% KMS solution for improving the dried product quality. For example, high contents of ascorbic acid and tannins, low acidity, and high scores for color and texture were observed when the KMS blanching was applied. These results suggest that KMS presents a protective effect on vitamin C and tannins, besides the well-known positive impact of KMS on color. Chinprahast et al. (2013) observed that blanching prior to vacuum impregnation was effective for producing minimally processed Indian gooseberry of good quality. They concluded that blanching in water at 90 °C for 60 s satisfactorily inhibited POD and produced a brighter and more yellow color in the product. Their results were attributed to inactivation of PPO, thus avoiding enzymatic browning, and to inactivation of POD, thus avoiding carotenoid destruction. In addition, blanching desirably reduced the astringency of Indian gooseberry by promoting the diffusion of tannins into the blanching water.

Nigerian okra seed (*Abelmoschus esculentus* Moench) was pretreated by soaking, blanching, malting, and roasting in a study by Adelakun et al. (2012). They found that blanching reduced the mineral content of the flour made from the seeds, increased its water and oil absorption capacity and its foam forming capacity, and reduced its emulsion ability/stability and its foam stability. They attributed such results to leaching, in the case of reduced mineral content; partial protein denaturation and starch gelatinization, in the case of increased water absorption capacity; and enhanced exposition of proteins to interact with oil due to micro-porous nature of the blanched product, in the case of increased oil absorption capacity.

The kinetics of color, texture, polyphenols content, and antioxidant capacity changes in York cabbage during blanching was elucidated by Jaiswal et al. (2012), who found that blanching promotes significant changes in the quality of cabbage. These changes could be well described by the zero-order model, the first-order model (Eq. 2.4), and the fractional conversion first-order model (Eq. 2.5), respectively. The former is shown below:

$$A = A_0 - kt \qquad (2.11)$$

where A is the parameter to be estimated, the subscript 0 indicates the initial value of the parameter, t is the blanching time, and k is the rate constant at temperature T. They observed that color (Fig. 2.12a) and texture (Fig. 2.12b) were affected until the last moments of blanching, while the phytochemical content (Figs. 2.13a and 2.13b) and the antioxidant activity (Fig. 2.14) suffered the most expressive decrease during the two first minutes of blanching. The effect of blanching on the degradation of the color property chroma (C^*) may be attributed to leaching or thermal degradation of water-soluble, heat-sensitive pigments. The same mechanisms are probably responsible for the decrease in phenolics content. The effect of HTST blanching on vegetables firmness decrease is well described too, and in the case of cabbage, that is poor in starch, could be due to the effect of heat on pectic substances.

Fig. 2.12 Experimental [80 (*multiplication sign*), 90 (*filled triangle*), and 100 °C (*white bullet*)] and predicted [80 (*en dash*), 90 (*em dash*), and 100 °C (*solid line*)] data for blanching temperature and time combinations on **a** color (chroma) (zero-order equation) and **b** texture (firmness) of York cabbage (first-order kinetics model) (Jaiswal et al. 2012)

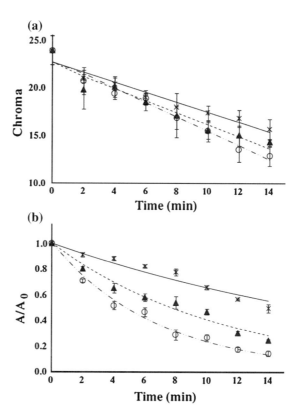

Herbs and spices can be blanched in order to increase their shelf life by means of inactivating undesirable enzymes. In the study of Kaiser et al. (2012), they found that water blanching prior to mincing of parsley was superior to mincing followed by heating in terms of chlorophylls preservation, pheophytin formation, and intensity of green color, but inferior with regard to phenolics preservation. In addition, their results showed that steam blanching was a good option for preserving phenolics, but not for color quality. They concluded that HTST water blanching was the best pretreatment for preparing paste-like parsley products free from undesirable enzymatic activity and of attractive color.

Lin et al. (2012) measured and modeled the POD and the PPO activity in leafs of *Rabdosia serra* (Maxim.) during blanching, finding that HTST water blanching (90 °C) or steam blanching (100 °C) for 90 s were the best strategies for the inactivation of these undesirable enzymes while retaining a high level of phenolics. They observed that the inactivation kinetics was well fit by the first-order model (Eq. 2.4) and its temperature dependence obeyed the Arrhenius model (Eq. 2.6), being the activation energy (E_a) for PPO inactivation higher than the activation energy for POD inactivation, which led to the conclusion that PPO should be the considered the target enzyme during blanching of *R. serra* leafs. Even though hot water blanching may promote thermal degradation, leaching, and diffusion of

Fig. 2.13 Experimental [80 (*multiplication sign*), 90 (*filled triangle*), and 100 °C (*white bullet*)] and predicted [80 (*en dash*), 90 (*em dash*), and 100 °C (*solid line*)] (fractional conversion first-order kinetics model) data for blanching temperature and time combinations on **a** total phenolic content and **b** total flavonoid content of York cabbage (Jaiswal et al. 2012)

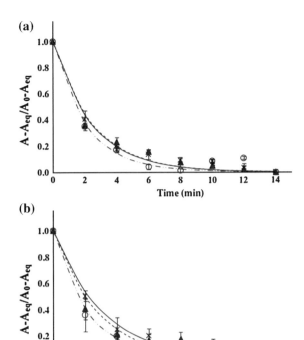

phenolic compounds, it improves the extractability of these compounds resulting in a roughly unaffected phenolic content after blanching.

Blanching of bell pepper in hot water inactivated lipoxygenase (LOX) in the tissue and consequently avoided the enzymatic generation of volatile compounds during frozen storage and thawing (Wampler and Barringer 2012). In addition, their results showed that the concentration of the majority of the volatile compounds did not change significantly during blanching.

Babiker and Eltoum (2013) found that blanching of tomatoes combined with gum arabic coating before air or sun drying improved the quality of the product as compared to unblanched/uncoated samples. They concluded that steam blanching for 3 min was especially effective for preventing browning, which was related to the inactivation of enzymes.

Inactivating PPO or minimizing its activity is a major goal during blanching of many vegetables that present a light-colored flesh, since this enzyme causes enzymatic browning that impairs the product visual quality. In the study by Goyeneche et al. (2013), the specificity of radish PPO for different substrates and

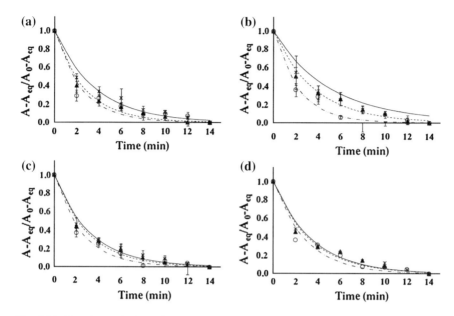

Fig. 2.14 Experimental [80 (*multiplication sign*), 90 (*filled triangle*), and 100 °C (*white bullet*)] and predicted [80 (*en dash*), 90 (*em dash*), and 100 °C (*solid line*)] (fractional conversion first-order kinetics model) data for blanching temperature and time combinations on antioxidant capacity on **a** DPPH radical scavenging capacity, **b** H_2O_2 radical scavenging capacity, **c** lipid peroxidation inhibitory ability, and **d** ferric reducing antioxidant potential of York cabbage (Jaiswal et al. 2012)

the effect of pH and temperature on its activity were investigated. They found that pyrocatechol, gallic acid, and pyrogallic acid were the substrates with higher specificity for radish PPO, and the effect of pH and temperature on enzymatic activity was substrate-dependent. When a blanching treatment was used, those researchers found that there are two forms of PPO in radish, namely a heat-labile and a heat-resistant form, which resulted in two clear zones in the curves of residual PPO activity versus blanching time for different blanching temperatures (Fig. 2.15). The inactivation kinetics was modeled using Eq. 2.7.

Koskiniemi et al. (2013) studied the effect of blanching pretreatment followed by different treatments with NaCl and citric acid on dielectric properties of sweet potato, broccoli, and Red Bell pepper, finding that blanching in water for 15 s at 90 °C was necessary for the reduction of pH to 3.8 in 24 h in sweet potato and broccoli equilibrated in a cover solution at room temperature. They attributed such behavior to the disruption of the cellular structure facilitating the transfer of NaCl and critic acid into the vegetables.

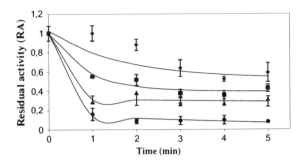

Fig. 2.15 Loss of activity for the radish PPO as function of blanching time and temperature (*n* = 3). Values are mean ± S.D., *n* = 3; *Filled diamond*, 60 °C; *Filled square*, 70 °C; *Filled triangle*, 80 °C; *Filled circle*, 90 °C; Eq. (2.7) (Goyeneche et al. 2013)

Ginger candy preparation was optimized toward product quality by means of a response surface design in which blanching time and product thickness were chosen as independent variables (Nath et al. 2013). They observed that higher blanching times resulted in lower hardness, lower soluble solids content, higher acidity, better taste, and higher overall acceptability. The results of the physical and chemical measurements may be attributed to the softening effect of heat, leaching, and higher penetration of the acid contained in the syrup, while the results of the sensory tests could indicate that the panelists prefer a softer and sourer ginger candy. They concluded that blanching of 10.9-mm-thick ginger slices in boiling water for 24.9 min, followed by dipping in hot aqueous citric acid/sugar solutions, was the best strategy for obtaining high-quality ginger candy.

Pectin was extracted from quince pomace by means of various process conditions in the study by Brown et al. (2014), who found that blanching at 70 °C for 10 min was superior to washing in cold water as a pretreatment. They optimized the process toward high yield, high level of galacturonic acid and high degree of methylation, concluding that the use of blanching was essential for achieving the desired quality goals.

The effect of blanching on color and bioactive compounds of asparagus, green beans, and zucchini was studied by Mazzeo et al. (2015). Their report shows that hot water blanching as performed for 30 s at 90 °C or for 2 min at 90 °C or for 2 min at 96 °C influenced the characteristics of asparagus, zucchini, and green beans, respectively. Color of the selected vegetables turned darker and greener after blanching, but these changes were not perfectly correlated to the changes in chlorophylls and pheophytin content. Color changes were attributed to air expulsion between the cells and filling of the remaining space with cell juice released through the damaged membranes. With regard to ascorbic acid and flavonoids, blanching was found to promote a low decrease and an increase in the content of these substances, respectively. In the case of ascorbic acid, leaching combined with enzymatic degradation and heat-driven destruction was claimed to promote losses, which were not higher because blanching inactivated ascorbate oxidase. With regard to flavonoids, the authors attributed their higher concentration in the blanched samples to a higher extractability after blanching.

Taro (*Colocasia esculenta*) was tentatively blanched in order to reduce its oil uptake during deep fat frying (Paz-Gamboa et al. 2015). This study showed that blanching of taro chips in hot water at 85 °C for 3 min caused a decrease of up to 80% in the amount of absorbed oil. Such behavior could be related to the gelatinization of starch, thus reducing oil absorption by the taro chips, as observed previously in other starchy systems, viz., potatoes (Reis et al. 2008).

Yacon (*Smallanthus sonchifolius*) is an Andean tuber recognized as a source of probiotic fibers, namely inulin and fructooligosaccharides. Scher et al. (2015) investigated the effect of hot water blanching on the levels of inulin, fructose, and glucose in yacon disks. They found that longer blanching times and higher blanching temperatures promoted higher losses of carbohydrates. This behavior may be attributed to increased diffusion and leaching of sugars due to heat-induced cell membrane disruption under severe blanching conditions. They concluded that yacon blanching must be performed at a temperature lower than 60 °C during 3 min or less, for maximal preservation of inulin.

The total phenolic and ascorbic content of beetroot, green pea, eggplant, and green pepper decreased with increasing blanching time and temperature following a linear pattern at 70 °C and a logarithmic pattern at 75–90 °C (Nambi et al. 2016). Furthermore, blanching promoted color changes and firmness loss, being these behaviors well described by zero- (Eq. 2.11) and first-order (Eq. 2.4) models, respectively.

Lotus rhizome slices subjected to blanching at 40 °C in 0.5% calcium chloride solution presented increased firmness as compared to unblanched samples and to samples blanched at 90 °C (Zhao et al. 2016). This result was attributed to the activation of PME along with the formation of calcium pectate in the tissue.

Final Considerations

In this chapter, relevant information on blanching and its impact on the physical, chemical, and sensory quality of several foods were provided. Potato, probably the most studied blanched food, was shown to be significantly affected by blanching, usually in a positive way. Studies on the blanching of carrots are also abundant and their results indicate that blanching is a suitable way for improving carrots quality. In addition, some reports showed that foods that are not usually blanched, like animal foods and fruits, can also have their quality improved by blanching. Nevertheless, a couple of reports on negative effects of blanching were also provided, showing that this process presents limitations. The overcoming of these limitations may be achieved by means of innovations in blanching technology, which are dealt with in Chap. 7.

References

Abiodun OA, Akinoso R (2015) Textural and sensory properties of trifoliate yam (Dioscorea dumetorum) flour and stiff dough "amala". J Food Sci Technol 52:2894–2901. Doi:10.1007/s13197-014-1313-y

Abu-Ghannam N, Crowley H (2006) The effect of low temperature blanching on the texture of whole processed new potatoes. J Food Eng 74:335–344. Doi:10.1016/j.jfoodeng.2005.03.025

Acevedo NC, Briones V, Buera P (2008) Microstructure affects the rate of chemical, physical and color changes during storage of dried apple discs. J Food Eng 85:222–231. Doi:10.1016/j.jfoodeng.2007.06.037

Adelakun OE, Ade-Omowaye BIO, Adeyemi IA et al (2012) Mineral composition and the functional attributes of Nigerian okra seed (Abelmoschus esculentus Moench) flour. Food Res Int 47:348–352. Doi:10.1016/j.foodres.2011.08.003

Afoakwa EO, Yenyi SE, Sakyi-Dawson E (2006) Response surface methodology for optimizing the pre-processing conditions during canning of a newly developed and promising cowpea (Vigna unguiculata) variety. J Food Eng 73:346–357. Doi:10.1016/j.jfoodeng.2005.01.037

Afoakwa EO, Budu AS, Merson AB (2007) Response surface methodology for studying the effect of processing conditions on some nutritional and textural properties of bambara groundnuts (Voandzei subterranea) during canning. Int J Food Sci Nutr 58:270–281. Doi:10.1080/09637480601154277

Agblor A, Scanlon MG (1998) Effects of blanching conditions on the mechanical properties of french fry strips. Am J Potato Res 75:245–255. Doi:10.1007/BF02853603

Agblor A, Scanlon MG (2000) Processing conditions influencing the physical properties of French fried potatoes. Potato Res 43:163–177. Doi:10.1007/BF02357957

Agüero MV, Ansorena MR, Roura SI et al (2008) Thermal inactivation of POD during blanching of butternut squash. LWT—Food Sci Technol 41:401–407. Doi:10.1016/j.lwt.2007.03.029

Aguilar CN, Anzaldua-Morales A, Talamas R et al (1997) Low-temperature blanch improves textural quality of French-fries. J Food Sci 62:568–571. Doi:10.1111/j.1365-2621.1997.tb04432.x

Al-Ruqaie IM (2006) Effect of different treatment processes and preservation methods on the quality of truffles: I. Conventional methods (drying/freezing). J Food Process Preserv 30:335–351

Altan A, McCarthy KL, Tikekar R et al (2011) Image analysis of microstructural changes in almond cotyledon as a result of processing. J Food Sci 76:E212–E221. Doi:10.1111/j.1750-3841.2010.01994.x

Alvarez MD, Morillo MJ, Canet W (2000) Characterization of the frying process of fresh and blanched potato strips using response surface methodology. Eur Food Res Technol 211:326–335. Doi:10.1007/s002170000161

Babiker EEB, Eltoum YAI (2013) Effect of edible surface coatings followed by dehydration on some quality attributes and antioxidants content of raw and blanched tomato slices. Food Sci Biotechnol 23:231–238. Doi:10.1007/s10068-014-0032-5

Balaswamy K, Rao PPG, Rao DG et al (2010) Effects of pretreatments and salt concentration on rohu (Labeo rohita) roes for preparation of roe pickle. J Food Sci Technol 47:219–223. Doi:10.1007/s13197-010-0035-z

Bechoff A, Wetsby A, Menya G et al (2010) Effect of pretreatments for retaining total carotenoids in dried and stored orange-fleshed-sweet potato chips. J Food Qual 34:259–267. Doi:10.1111/j.1745-4557.2011.00391.x

Bernaś E, Jaworska G (2016) Vitamins profile as an indicator of the quality of frozen Agaricus bisporus mushrooms. J Food Compos Anal 49:1–8. Doi:10.1016/j.jfca.2016.03.002

Brown VA, Lozano JE, Genovese DB (2014) Pectin extraction from quince (Cydonia oblonga) pomace applying alternative methods: effect of process variables and preliminary optimization. Food Sci Technol Int 20:83–98. Doi:10.1177/1082013212469616

Burande RR, Kumbhar BK, Ghosh PK et al (2008) Optimization of fluidized bed drying process of green peas using response surface methodology. Dry Technol 26:920–930. Doi:10.1080/07373930802142739

Califano AN, Calvelo A (1987) Adjustment of surface concentration of reducing sugars before frying of potato strips. J Food Process Preserv 12:1–9

Cam S, Kilic M (2009) Effect of blanching on storage stability of hazelnut meal. J Food Qual 32:369–380. Doi:10.1111/j.1745-4557.2009.00254.x

Carbonell S, Oliveira JC, Kelly AL (2006) Effect of pretreatments and freezing rate on the firmness of potato tissue after a freeze-thaw cycle. Int J Food Sci Technol 41:757–767. Doi:10.1111/j.1365-2621.2005.01054.x

Carillo P, Cacace D, de Rosa M et al (2009) Process optimisation and physicochemical characterisation of potato powder. Int J Food Sci Technol 44:145–151. Doi:10.1111/j.1365-2621.2007.01696.x

Christiaens S, Mbong VB, Van Buggenhout S et al (2012) Influence of processing on the pectin structure–function relationship in broccoli purée. Innov Food Sci Emerg Technol 15:57–65. Doi:10.1016/j.ifset.2012.02.011

Chinprahast N, Siripatrawan U, Leerahawong A et al (2013) Effects of blanching and vacuum impregnation on physicochemical and sensory properties of Indian gooseberry (*Phyllanthus emblica* L.). J Food Process Preserv 37:57–65. Doi:10.1111/j.1745-4549.2011.00613.x

Corriadini MG, Pelcg M (2008) Solid food foams. In: Aguilera JM, Lillford PJ (eds) Food materials science, 1st edn. Springer, Heidelberg, pp 169–202

Cunningham SE, Mcminn WAM, Magee TRA et al (2008) Effect of processing conditions on the water absorption and texture kinetics of potato. J Food Eng 84:214–223. Doi:10.1016/j.jfoodeng.2007.05.007

De Jesus RS, Lessi E, Tenuta-Filho A (2001) Estabilidade Química E Microbiológica De "Minced Fish" De Peixes Amazônicos Durante O Congelamento. Ciência e Tecnol Aliment 21:144–148. Doi:10.1590/S0101-20612001000200004

Durán M, Pedreschi F, Moyano P et al (2007) Oil partition in pre-treated potato slices during frying and cooling. J Food Eng 81:257–265. Doi:10.1016/j.jfoodeng.2006.11.004

Dutta D, Chaudhuri UR, Chakraborty R (2009) Degradation of total carotenoids and texture in frozen pumpkins when kept for storage under varying conditions of time and temperature. Int J Food Sci Nutr 60(Suppl 1):17–26. Doi:10.1080/09637480701850220

Eissa HA, Fouad GM, Shouk AEA (2009) Effect of some thermal and chemical pre-treatments on smoked oyster mushroom quality. Int J Food Sci Technol 44:251–261. Doi:10.1111/j.1365-2621.2007.01671.x

Fante L, Noreña CPZ (2012) Enzyme inactivation kinetics and colour changes in Garlic (*Allium sativum* L.) blanched under different conditions. J Food Eng 108:436–443. Doi:10.1016/j.jfoodeng.2011.08.024

Fernández C, Dolores Alvarez M, Canet W (2006) The effect of low-temperature blanching on the quality of fresh and frozen/thawed mashed potatoes. Int J Food Sci Technol 41:577–595. Doi:10.1111/j.1365-2621.2005.01119.x

Fernández-Segovia I, Escriche I, Gómez-Sintes M et al (2006) Influence of different preservation treatments on the volatile fraction of desalted cod. Food Chem 98:473–482. Doi:10.1016/j.foodchem.2005.06.021

Fernández-Segovia I, Escriche I, Fuentes A et al (2007) Microbial and sensory changes during refrigerated storage of desalted cod (*Gadus morhua*) preserved by combined methods. Int J Food Microbiol 116:64–72. Doi:10.1016/j.ijfoodmicro.2006.12.026

Gökmen V, Savaş Bahçeci K, Serpen A, Acar J (2005) Study of lipoxygenase and POD as blanching indicator enzymes in peas: change of enzyme activity, ascorbic acid and chlorophylls during frozen storage. LWT—Food Sci Technol 38:903–908. Doi:10.1016/j.lwt.2004.06.018

Gonçalves EM, Pinheiro J, Abreu M et al (2007) Modelling the kinetics of POD inactivation, colour and texture changes of pumpkin (*Cucurbita maxima* L.) during blanching. J Food Eng 81:693–701. Doi:10.1016/j.jfoodeng.2007.01.011

Gonçalves EM, Pinheiro J, Abreu M et al (2010) Carrot (*Daucus carota* L.) POD inactivation, phenolic content and physical changes kinetics due to blanching. J Food Eng 97:574–581. Doi:10.1016/j.jfoodeng.2009.12.005

Gonçalves EM, Abreu M, Brandão TRS et al (2011a) Degradation kinetics of colour, vitamin C and drip loss in frozen broccoli (*Brassica oleracea* L. ssp. Italica) during storage at isothermal and non-isothermal conditions. Int J Refrig 34:2136–2144. Doi:10.1016/j.ijrefrig.2011.06.006

Gonçalves EM, Pinheiro J, Abreu M et al (2011b) Kinetics of quality changes of pumpkin (*Curcurbita maxima* L.) stored under isothermal and non-isothermal frozen conditions. J Food Eng 106:40–47. Doi:10.1016/j.jfoodeng.2011.04.004

González-Martínez G, Ahrné L, Gekas V et al (2004) Analysis of temperature distribution in potato tissue during blanching and its effect on the absolute residual pectin methylesterase activity. J Food Eng 65:433–441. Doi:10.1016/j.jfoodeng.2004.02.003

Gowen A, Abu-Ghannam N, Frias J et al (2007) Influence of pre-blanching on the water absorption kinetics of soybeans. J Food Eng 78:965–971. Doi:10.1016/j.jfoodeng.2005.12.009

Goyeneche R, Di Scala K, Roura S (2013) Biochemical characterization and thermal inactivation of polyphenol oxidase from radish (*Raphanus sativus* var. sativus). LWT—Food Sci Technol 54:57–62. Doi:10.1016/j.lwt.2013.04.014

He J, Cheng L, Gu Z et al (2014) Effects of low-temperature blanching on tissue firmness and cell wall strengthening during sweet potato flour processing. Int J Food Sci Technol 49:1360–1366. Doi:10.1111/ijfs.12437

Ismail N, Revathi R (2006) Studies on the effects of blanching time, evaporation time, temperature and hydrocolloid on physical properties of chili (*Capsicum annum* var kulai) puree. LWT—Food Sci Technol 39:91–97. Doi:10.1016/j.lwt.2004.12.003

Jaiswal AK, Gupta S, Abu-Ghannam N (2012) Kinetic evaluation of colour, texture, polyphenols and antioxidant capacity of Irish York cabbage after blanching treatment. Food Chem 131:63–72. Doi:10.1016/j.foodchem.2011.08.032

Jaworska G, Bernaś E, Biernacka A et al (2010) Comparison of the texture of fresh and preserved *Agaricus bisporus* and *Boletus edulis* mushrooms. Int J Food Sci Technol 45:1659–1665. Doi:10.1111/j.1365-2621.2010.02319.x

Jaworska G, Bernaś E, Mickowska B (2011) Effect of production process on the amino acid content of frozen and canned *Pleurotus ostreatus* mushrooms. Food Chem 125:936–943. Doi:10.1016/j.foodchem.2010.09.084

Jin X, van der Sman RGM, van Maanen JFC et al (2014) Moisture sorption isotherms of broccoli interpreted with the Flory-Huggins free volume theory. Food Biophys 9:1–9. Doi:10.1007/s11483-013-9311-6

Kadam DM, Samuel DVK, Chandra P et al (2008) Impact of processing treatments and packaging material on some properties of stored dehydrated cauliflower. Int J Food Sci Technol 43:1–14. Doi:10.1111/j.1365-2621.2006.01372.x

Kaiser A, Brinkmann M, Carle R et al (2012) Influence of thermal treatment on color, enzyme activities, and antioxidant capacity of innovative pastelike parsley products. J Agric Food Chem 60:3291–3301. Doi:10.1021/jf205098q

Kmiecik W, Lisiewska Z, Słupski J et al (2008) Effect of preliminary and culinary processing on amino acid content and protein quality in frozen French beans. Int J Food Sci Technol 43:1786–1791. Doi:10.1111/j.1365-2621.2007.01702.x

Koca N, Karadeniz F, Burdurlu HS (2007) Effect of pH on chlorophyll degradation and colour loss in blanched green peas. Food Chem 100:609–615. Doi:10.1016/j.foodchem.2005.09.079

Koskiniemi CB, Truong V-D, McFeeters RF et al (2013) Effects of acid, salt, and soaking time on the dielectric properties of acidified vegetables. Int J Food Prop 16:917–927. Doi:10.1080/10942912.2011.567428

Kotwaliwale N, Bakane P, Verma A (2007) Changes in textural and optical properties of oyster mushroom during hot air drying. J Food Eng 78:1207–1211. Doi:10.1016/j.jfoodeng.2005.12.033

Kulkarni SG, Vijayanand P, Aksha M et al (2008) Effect of dehydration on the quality and storage stability of immature dates (*Pheonix dactylifera*). LWT—Food Sci Technol 41:278–283. Doi:10.1016/j.lwt.2007.02.023

Lavelli V, Caronni P (2010) Polyphenol oxidase activity and implications on the quality of intermediate moisture and dried apples. Eur Food Res Technol 231:93–100. Doi:10.1007/s00217-010-1256-0

Lemmens L, Colle I, Knockaert G et al (2013) Influence of pilot scale in pack pasteurization and sterilization treatments on nutritional and textural characteristics of carrot pieces. Food Res Int 50:526–533. Doi:10.1016/j.foodres.2011.02.030

Lespinard AR, Goñi SM, Salgado PR et al (2009) Experimental determination and modelling of size variation, heat transfer and quality indexes during mushroom blanching. J Food Eng 92:8–17. Doi:10.1016/j.jfoodeng.2008.10.025

Lewicki PP, Pawlak G (2003) Effect of drying on microstructure of plant tissue. Dry Technol 21:657–683. Doi:10.1081/DRT-120019057

Lin L, Lei F, Sun D et al (2012) Thermal inactivation kinetics of Rabdosia serra (Maxim.) Hara leaf POD and polyphenol oxidase and comparative evaluation of drying methods on leaf phenolic profile and bioactivities. Food Chem 134:2021–2029. Doi:10.1016/j.foodchem.2012.04.008

Ling AC, Lund DB (1978) Determining kinetic parameters for thermal inactivation of heat-resistant and heat-labile isozymes from thermal destruction curves. J Food Sci 43:1307–1310

Lisiewska Z, Korus A, Kmiecik W et al (2006) Effect of maturity stage on the content of ash components in raw and preserved grass pea (*Lathyrus sativus* L.) seeds. Int J Food Sci Nutr 57:39–45. Doi:10.1080/09637480500515420

Lisiewska Z, Slupski J, Kmiecik W et al (2008) Availability of essential and trace elements in frozen leguminous vegetables prepared for consumption according to the method of pre-freezing processing. Food Chem 106:576–582. Doi:10.1016/j.foodchem.2007.06.025

Lisiewska Z, Kmiecik W, Gębczyński P et al (2011) Amino acid profile of raw and as-eaten products of spinach (*Spinacia oleracea* L.). Food Chem 126:460–465. Doi:10.1016/j.foodchem.2010.11.015

Lisińska G, Gołubowska G (2005) Structural changes of potato tissue during French fries production. Food Chem 93:681–687. Doi:10.1016/j.foodchem.2004.10.046

Liu EZ, Scanlon MG (2007) Modeling the effect of blanching conditions on the texture of potato strips. J Food Eng 81:292–297. Doi:10.1016/j.jfoodeng.2006.08.002

Liu F, Grauwet T, Kebede BT et al (2014) Comparing the effects of high hydrostatic pressure and thermal processing on blanched and unblanched mango (*Mangifera indica* L.) nectar: using headspace fingerprinting as an untargeted approach. Food Bioprocess Technol 1–12. Doi:10.1007/s11947-014-1280-3

López-Malo A, Palou E (2008) Storage stability of pineapple slices preserved by combined methods. Int J Food Sci Technol 43:289–295. Doi:10.1111/j.1365-2621.2006.01433.x

Lv Y-C, Song H-L, Li X et al (2011) Influence of blanching and grinding process with hot water on beany and non-beany flavor in soymilk. J Food Sci 76:S20–S25. Doi:10.1111/j.1750-3841.2010.01947.x

Ma T, Luo J, Tian C et al (2015) Influence of technical processing units on chemical composition and antimicrobial activity of carrot (*Daucus carrot* L.) juice essential oil. Food Chem 170:394–400. Doi:10.1016/j.foodchem.2014.08.018

Mazzeo T, Paciulli M, Chiavaro E et al (2015) The impact of the industrial freezing process on selected vegetables Part II. Colour and bioactive compounds. Food Res Int 75:89–97. Doi:10.1016/j.foodres.2015.05.036

Montouto-Graña M, Cabanas-Arias S, Vázquez-Odériz ML et al (2011) Industrially processed vacuum-packed peeled kennebec potatoes: process optimization, sensory evaluation, and consumer response. J Food Sci 76:S314–S318. Doi:10.1111/j.1750-3841.2011.02185.x

Moon C-Y, Yoon W-B, Hahm Y-T et al (2011) Optimization of processing conditions and evaluation of shelf-life for jeonbokjang products. Food Sci Biotechnol 20:1419–1424. Doi:10. 1007/s10068-011-0195-2

Moyano PC, Berna AZ (2002) Modeling water loss during frying of potato strips: effect of solute impregnation. Dry Technol 20:1303

Moyano PC, Troncoso E, Pedreschi F (2007) Modeling texture kinetics during thermal processing of potato products. J Food Sci 72:102–107. Doi:10.1111/j.1750-3841.2006.00267.x

Mozzoni LA, Chen P, Morawicki RO et al (2009) Quality attributes of vegetable soybean as a function of boiling time and condition. Int J Food Sci Technol 44:2089–2099. Doi:10.1111/j. 1365-2621.2009.02038.x

Murugkar DA (2014) Effect of different process parameters on the quality of soymilk and tofu from sprouted soybean. J Food Sci Technol 52:2886–2893. Doi:10.1007/s13197-014-1320-z

Nambi VE, Gupta RK, Kumar S et al (2016) Degradation kinetics of bioactive components, antioxidant activity, colour and textural properties of selected vegetables during blanching. J Food Sci Technol. Doi:10.1007/s13197-016-2280-2

Nath A, Deka BC, Jha AK et al (2013) Effect of slice thickness and blanching time on different quality attributes of instant ginger candy. J Food Sci Technol 50:197–202. Doi:10.1007/ s13197-012-0619-x

Ndangui CB, Petit J, Gaiani C et al (2014) Impact of thermal and chemical pretreatments on physicochemical, rheological, and functional properties of sweet potato (*Ipomea batatas* Lam) flour. Food Bioprocess Technol 7:3618–3628. Doi:10.1007/s11947-014-1361-3

Ndiaye C, Xu SY, Wang Z (2009) Steam blanching effect on polyphenoloxidase, POD and colour of mango (*Mangifera indica* L.) slices. Food Chem 113:92–95. Doi:10.1016/j.foodchem.2008. 07.027

Neri L, Hernando IH, Pérez-Munuera I et al (2011) Effect of blanching in water and sugar solutions on texture and microstructure of sliced carrots. J Food Sci 76:E23–E30. Doi:10.1111/ j.1750-3841.2010.01906.x

Neri L, Hernando I, Pérez-Munuera I et al (2014) Mechanical properties and microstructure of frozen carrots during storage as affected by blanching in water and sugar solutions. Food Chem 144:65–73. Doi:10.1016/j.foodchem.2013.07.123

Oner ME, Walker PN (2011) Shelf-life of near-aseptically packaged refrigerated potato strips. LWT—Food Sci Technol 44:1616–1620. Doi:10.1016/j.lwt.2011.02.003

Oner ME, Walker PN, Demirci A (2011) Effect of in-package gaseous ozone treatment on shelf life of blanched potato strips during refrigerated storage. Int J Food Sci Technol 46:406–412. Doi:10.1111/j.1365-2621.2010.02503.x

Paciulli M, Ganino T, Carini E et al (2016) Effect of different cooking methods on structure and quality of industrially frozen carrots. J Food Sci Technol 53:2443–2451. Doi:10.1007/s13197-016-2229-5

Paz-Gamboa E, Ramírez-Figueroa E, Vivar-Vera MA et al (2015) Study of oil uptake during deep-fat frying of Taro (*Colocasia esculenta*) chips. CyTA—J Food 6337:1–6. Doi:10.1080/ 19476337.2015.1010587

Pedreschi F, Hernandez P, Figueroa C, Moyano P (2005) Modeling water loss during frying of potato slices. Int J Food Prop 8:289–299. Doi:10.1081/JFP-200059480

Pedreschi F, Moyano P, Santis N, Pedreschi R (2007) Physical properties of pre-treated potato chips. J Food Eng 79:1474–1482. Doi:10.1016/j.jfoodeng.2006.04.029

Pedreschi F, Travisany X, Reyes C et al (2009) Kinetics of extraction of reducing sugar during blanching of potato slices. J Food Eng 91:443–447. Doi:10.1016/j.jfoodeng.2008.09.022

Peng X, Guo S (2014) Texture characteristics of soymilk gels formed by lactic fermentation: a comparison of soymilk prepared by blanching soybeans under different temperatures. Food Hydrocoll 43:58–65. Doi:10.1016/j.foodhyd.2014.04.034

Prajapati VK, Nema PK, Rathore SS (2011) Effect of pretreatment and drying methods on quality of value-added dried aonla (*Emblica officinalis* Gaertn) shreds. J Food Sci Technol 48:45–52. Doi:10.1007/s13197-010-0124-z

Reis FR, Masson ML, Waszczynskyj N (2008) Influence of a blanching pretreatment on color, oil uptake and water activity of potato sticks, and its optimization. J Food Process Eng 31:833–852. Doi:10.1111/j.1745-4530.2007.00193.x

Rejano L, Sánchez AH, Montaño A et al (2007) Kinetics of heat penetration and textural changes in garlic during blanching. J Food Eng 78:465–471. Doi:10.1016/j.jfoodeng.2005.10.016

Rimac-Brnčić S, Lelas V, Rade D, Šimundić B (2004) Decreasing of oil absorption in potato strips during deep fat frying. J Food Eng 64:237–241. Doi:10.1016/j.jfoodeng.2003.10.006

Roos YH (2008) The Glassy State. In: Aguilera JM, Lillford PJ (eds) Food materials science, 1st edn. Springer, Heidelberg, pp 67–82

Rudra Shalini G, Shivhare US, Basu S (2008) Thermal inactivation kinetics of POD in mint leaves. J Food Eng 85:147–153. Doi:10.1016/j.jfoodeng.2007.07.010

Santis N, Mendoza F, Moyano P et al (2007) Soaking in a NaCl solution produce paler potato chips. LWT—Food Sci Technol 40:307–312. Doi:10.1016/j.lwt.2005.09.020

Sarang S, Sastry SK, Gaines J et al (2007) Product formulation for ohmic heating: blanching as a pretreatment method to improve uniformity in heating of solid-liquid food mixtures. J Food Sci 72:E227–E234. Doi:10.1111/j.1750-3841.2007.00380.x

Scher CF, Brandelli A, Noreña CZ (2015) Yacon inulin leaching during hot water blanching. Ciência e Agrotecnologia 39:523–529. Doi:10.1590/S1413-70542015000500011

Schweiggert U, Mix K, Schieber A, Carle R (2005) An innovative process for the production of spices through immediate thermal treatment of the plant material. Innov Food Sci Emerg Technol 6:143–153. Doi:10.1016/j.ifset.2004.11.006

Schweiggert U, Schieber A, Carle R (2006) Effects of blanching and storage on capsaicinoid stability and POD activity of hot chili peppers (*Capsicum frutescens* L.). Innov Food Sci Emerg Technol 7:217–224. Doi:10.1016/j.ifset.2006.03.003

Shivhare US, Gupta M, Basu S et al (2009) Optimization of blanching process for carrots. J Food Process Eng 32:587–605. Doi:10.1111/j.1745-4530.2007.00234.x

Singh G, Sehgal S, Kawatra A (2006) Mineral profile, anti-nutrients and in vitro digestibility of biscuit prepared from blanched and malted pearl millet flour. Nutr Food Sci 36:231–239. Doi:10.1108/00346650610676802

Słupski J (2010) Evaluation of the amino acid content and sensory value of flageolet bean seeds (*Phaseolus vulgaris* L.) as affected by preprocessing methods before freezing. Int J Food Sci Technol 45:1068–1075. Doi:10.1111/j.1365-2621.2010.02240.x

Słupski J, Achrem-Achremowicz J, Lisiewska Z et al (2010a) Effect of processing on the amino acid content of New Zealand spinach (*Tetragonia tetragonioides* Pall. Kuntze). Int J Food Sci Technol 45:1682–1688. Doi:10.1111/j.1365-2621.2010.02315.x

Słupski J, Korus A, Lisiewska Z et al (2010b) Content of amino acids and the quality of protein in as-eaten green asparagus (*Asparagus officinalis* L.) products. Int J Food Sci Technol 45:733–739. Doi:10.1111/j.1365-2621.2010.02193.x

Sobukola OP, Awonorin SO, Sanni LO et al (2008) Optimization of blanching conditions prior to deep fat frying of yam slices. Int J Food Prop 11:379–391. Doi:10.1080/10942910701409294

Sreedhar RV, Roohie K, Venkatachalam L et al (2007) Specific pretreatments reduce curing period of vanilla (*Vanilla planifolia*) beans. J Agric Food Chem 55:2947–2955. Doi:10.1021/jf063523k

Sreenath PG, Abhilash S, Ravishankar CN et al (2008) Standardization of process parameters for ready-to-eat shrimp curry in tin-free steel cans. J Food Process Preserv 32:247–269

Taiwo KA, Baik OD (2007) Effects of pre-treatments on the shrinkage and textural properties of fried sweet potatoes. LWT—Food Sci Technol 40:661–668. Doi:10.1016/j.lwt.2006.03.005

Tansey F, Gormley R, Butler F (2010) The effect of freezing compared with chilling on selected physico-chemical and sensory properties of sous vide cooked carrots. Innov Food Sci Emerg Technol 11:137–145. Doi:10.1016/j.ifset.2009.11.001

Tijskens LMM, Waldron KW, Ng A et al (1997) The kinetics of PME in potatoes and carrots during blanching. J Food Eng 34:371–385. Doi:10.1016/S0260-8774(98)00005-3

Utomo JS, Man YBC, Rahman RA et al (2008) The effect of shape, blanching methods and flour on characteristics of restructured sweetpotato stick. 1896–1900. Doi:10.1111/j.1365-2621. 2008.01792.x

Van Dyk S, McGlasson WB, Williams M et al (2010) Influence of curing procedures on sensory quality of vanilla beans. Fruits 65:387–399. Doi:10.1051/fruits/2010033

Varnalis AI, Brennan JG, MacDougall DB (2001a) Proposed mechanism of high temperature puffing of potato. Part II. Influence of blanching and initial drying on the permeability of the partially dried layer to water vapour. J Food Eng 48:369–378. Doi:10.1016/S0260-8774(00) 00198-9

Varnalis AI, Brennan JG, MacDougall DB (2001b) Proposed mechanism of high-temperature puffing of potato. Part I. The influence of blanching and drying conditions on the volume of puffed cubes. J Food Eng 48:361–367. Doi:10.1016/S0260-8774(00)00197-7

Verlinden B, Yuksel D, Baheri M et al (2000) Low temperature blanching effect on the changes in mechanical properties during subsequentcooking of three potato cultivars. Int J Food Sci Technol 35:331–340

Waisundara VY, Perera CO, Barlow PJ (2007) Effect of different pre-treatments of fresh coconut kernels on some of the quality attributes of the coconut milk extracted. Food Chem 101:771–777. Doi:10.1016/j.foodchem.2006.02.032

Wallerstein JS, Ralph Thomas A, Hale MG (1947) Studies on oxidase activity in potato tubers I. o-phenylenediamine as a colorimetric reagent. BBA—Biochim Biophys Acta 1:175–183

Wampler B, Barringer SA (2012) Volatile generation in bell peppers during frozen storage and thawing using selected ion flow tube mass spectrometry (SIFT-MS). J Food Sci 77:C677–C683. Doi:10.1111/j.1750-3841.2012.02727.x

Wang W, Sastry S (1997) Changes in electrical conductivity of selected vegetables during multiple thermal treatments. J Food Process Eng 20:499–516

Wennberg M, Ekvall J, Olsson K et al (2006) Changes in carbohydrate and glucosinolate composition in white cabbage (*Brassica oleracea* var. capitata) during blanching and treatment with acetic acid. Food Chem 95:226–236. Doi:10.1016/j.foodchem.2004.11.057

Whitaker JR (1996) Enzymes. In: Fennema OR (ed) Food chemistry, 3rd edn. Marcel Dekker, New York

Yu LJ, Rupasinghe HPV (2013) Improvement of cloud stability, yield and β-carotene content of carrot juice by process modification. Food Sci Technol Int 19:399–406. Doi:10.1177/1082013212455342

Zhao W, Xie W, Du S et al (2016) Changes in physicochemical properties related to the texture of lotus rhizomes subjected to heat blanching and calcium immersion. Food Chem 211:409–414. Doi:10.1016/j.foodchem.2016.05.075

Zorzella CA, Vendruscolo JLS, Treptow RO et al (2003) Physical, chemical and sensory characterization of different potato genotypes, processed in the form of chips. Braz J Food Technol 6:15–24

Chapter 3
Effect of Blanching on Food Bioactive Compounds

Bogdan Demczuk Junior

Abstract Foods like fruit, vegetables, and whole grains contain expressive amounts of bioactive phytochemicals. They can provide important health benefits like the decrease in the risk of chronical diseases. Bioactive compounds include carotenoids, tocopherols, and ascorbic acid. In addition, the greatest antioxidant effect related to vegetable matrices is ascribed to phenolic compounds. Such compounds can also be associated with flavor and aroma of fruit and vegetables. Despite the fact that heat treatments used for cooking are important to provide the food with sensory acceptability and increased digestibility, they can cause undesirable decrease in micronutrients and bioactive compounds content. Therefore, the objective of the present chapter is to deal with the influence of blanching on food quality as related to bioactive compounds. Among the most important topics that will be dealt with are the transformations involving carotenoids, phenolic compounds, antioxidant activity, organosulfur compounds, and fructooligosaccharides.

Keywords Blanching · Bioactive compounds · Ascorbic acid · Phenolic compounds · Antioxidant activity

Impact of Blanching on the Carotenoid Content of Foods

Carotenoids comprise a wide group of liposoluble compounds that are widespread in nature. In food, carotenoids can be found in the form of tetraterpenoids of 40 carbons bound by opposed units in the center of the molecule. The conjugated system of double bounds in the chain is responsible for the light absorption capacity of carotenoids, providing colors that range from yellow to red in fruit and vegetables (Rodriguez-Amaya 2001; Rodriguez-Amaya and Kimura 2004).

B. Demczuk Junior (✉)
Food Engineering Course, Federal Technological University of Paraná—Campus Campo Mourão, Campo Mourão, Paraná, Brazil
e-mail: bdjunior@gmail.com

© Springer International Publishing AG 2017
F. Richter Reis (ed.), *New Perspectives on Food Blanching*,
DOI 10.1007/978-3-319-48665-9_3

Among the carotenoids regarded as beneficial for human health are β-carotene, α-carotene, β-cryptoxanthin, lutein e lycopene. Besides acting as precursors of vitamin A that is fundamental for preventing macular degeneration and cataract, carotenoids seem to be related to other health effects like increase in immune response, decrease in the risk of cardiovascular and degenerative diseases, like cancer (Rodriguez-Amaya and Kimura 2004; Uenojo et al. 2007).

Carotenoid content in foods is not severely affected by household cooking methods, like microwaves, steam, or boiling. On the other hand, severe heating may cause carotenoid destruction by oxidation. For example, carotenoid losses are observed during stir frying due to solubilization. Therefore, it is important to process food in a way that preserves its carotenoids.

According to Mayer-Miebach and Spieß (2003), different effects of heat treatments like blanching, cooking, frying and boiling, on carotenoids will be observed depending on the time and the temperature used.

During mild heat treatments like blanching, pasteurization and cooking, in which temperatures range from 60 to 100 °C, great part of carotenoids is retained and isomerization is insignificant. At the same time, release of bound carotenoids by means of tissue rupture increases their bioavailability (Maiani et al. 2009).

One of the advantages of blanching is the inactivation of oxidative enzymes, preventing future losses during processing (Rodriguez-Amaya and Kimura 2004). On the other hand, significant losses are reported during processing at very high temperatures, like during frying (Amoussa-Hounkpatin et al. 2013).

Green Leafy Vegetables

Foods of animal origin that contain vitamin A, like dairy, eggs and liver, are seldom consumed by low-income populations. On the other hand, green leafy vegetables harvested and consumed in western Africa may be rich in beta-carotene (Veda et al. 2010).

Amaranth leaves (*Amaranthus cruentus*) found in regions like Africa and Asia are an example of a cheap source of carotenoids (Adebooye et al. 2008). Amoussa-Hounkpatin et al. (2013) found that blanching at about 100 °C for 18 min did not affect the carotenoid profile or the retinol activity equivalents of amaranth leaves. However, the measurement of 9-cis-beta-carotene and 13-cis-beta-carotene showed that these compounds were present in higher levels in cooked leaves when compared to fresh leaves. The authors suggest that this result may be attributed to isomerization reaction during heat treatment. The same report showed that the use of alkaline additives typically used in the region of the study prevented carotenoid degradation, suggesting that these compounds are more stable than hydro soluble vitamins under alkaline conditions.

Amoussa-Hounkpatin et al. (2013) found that violaxanthin was the only carotenoid affected during heat treatment of amaranth leaves in their study. The values of retinol activity equivalents remained high after blanching, even when the process was performed in alkaline medium.

Bunea et al. (2008) obtained a low decrease (about 15%) in the total carotenoid content of spinach blanched before freezing after one month of storage. Higher losses (about 64%) were detected in blanched spinach before storage under refrigeration for 24–72 h. The authors also assessed the content of selected carotenoids, finding that lutein and beta-carotene decreased more than 50% when stored at 4 °C. However, when blanched, spinach retained these carotenoids after a month of frozen storage.

Cruciferous Vegetables

Ahmed and Ali (2013) studied the effect of processing on the features of fresh cauliflower containing an initial carotenoid content of 126.22 mg/100 g dry matter. There was no significant difference between carotenoid content of steam blanched (3 min) and steam cooked (6 min 15 s) cauliflower, which were higher than hot water blanched (3 min) and microwave cooked (1000 W/3.5 min) cauliflower. Hot water cooked and (6 min) and stir fried (140 °C/4.5 min) samples presented the lowest carotenoid content.

Bernhardt and Schlich (2006) related the cell membrane rupture caused by blanching to significant all-trans-β-carotene losses in broccoli.

Roots

Chantaro et al. (2008) obtained antioxidant high dietary fiber powder from carrot peels, a byproduct of the carrot processing industry. The authors evaluated the effect of blanching and hot air drying on the features of carrot peel flour, including beta-carotene content. Results showed that blanching presented a significant effect on the content of fiber and other compounds of carrot peels, and also on water retention capacity and swelling capacity of dietary fiber powder. Thermal degradation during blanching and drying caused decrease in beta-carotene and phenolic compounds content, thereby reducing the antioxidant capacity of the final product.

Impact of Blanching on the Phenolic Content of Foods

Phenolic compounds can be found in fruit and vegetables either in soluble or combined forms, i.e., in the form of complexes with the cell wall. They contribute directly to the antioxidant activity of these systems.

The high surface area of vegetables in contact with water and the high processing temperatures may be considered the probable causes of the damages to cell
wall and phenolic compounds that take place during heat treatments like cooking
and blanching. Leaching can also occur, since the majority of phenolic compounds
are water-soluble (Zhang and Hamauzu 2004; Francisco et al. 2010).

Flavonoids are phenolics compounds whose concentration in food depends on
their sensitivity to modification or degradation, on the processing method and on
the processing time. Reports on the effect of heat treatment in aqueous solution on
flavonoid content of foods show that different flavonoids present distinct sensitiv-
ities to degradation depending on the structural variability of the so-called C ring
(Bernaert et al. 2014).

Cruciferous Vegetables

Ahmed and Ali (2013) found total phenolic content of 782.43 mg/100 g on dry
basis in fresh cauliflower. After blanching and other heat treatments significant
losses in phenolic compounds ranging from 15.6 and 51.9% were detected. For the
sake of comparison, the lowest losses were observed for steam blanching (3 min)
and steam boiling (6.25 min), followed by stir frying (140 °C/4.5 min) and
microwave cooking (1000 W/3.5 min). The highest losses of total phenolics were
observed for water blanching for 3 min (37.7%) and water boiling for 6 min
(51.9%). With regard to flavonoids, the highest losses were observed during water
boiling (56.39%), water blanching (43.42%) and stir frying (30.23%).

Porter (2012), when conducting a study on processing time and processing
method (boiling in water or microwave), found that boiling for 5 min promoted a
49.55% decrease in flavonoid content of purple sprouting broccoli.

Fruits

Del Bo et al. (2012) investigated the absorption of anthocyanins after consumption
of 300 g of blueberry puree obtained from blanched and unblanched fruits. They
found that the blood concentration of phenolics in healthy humans was higher after
consumption of blanched blueberry puree in detriment of unblanched puree. The
maximal absorption of anthocyanins in blood plasma was observed after 1.5 h of
ingestion of the puree, being significantly higher ($p < 0.05$) after the ingestion of
blanched blueberry puree. They concluded that blanching did not affect the
anthocyanin content of the purees, yet it promoted the increase in anthocyanin
absorption by humans who ingested the product.

Brambilla et al. (2008) used blanching as a pretreatment during blueberry juice
processing and assessed some of the product features, finding that blanching has a
protective effect on blueberry anthocyanins, yet each anthocyanin respond

differently to processing. Their conclusions suggest that the anthocyanin profile of blueberry juice and raw blueberry are more similar when fruits are blanched before juicing.

Fang et al. (2006) studied the effect of processing on the quality of bayberry juice. Samples comprised juices made from: untreated fruit; SO_2 treated fruit; pasteurized fruit puree; and blanched fruit. The author found that losses of polyphenols due to processing ranged from 57 to 74%. The polyphenolic content of blanched and pasteurized samples was significantly higher ($p < 0.05$) than control and SO_2 treated samples. Additionally, the heat treated juices presented higher anthocyanin content than SO_2 treated samples. They affirm that the heat may have inactivated PPO avoiding phenolic degradation and also ruptured the cell membrane thus allowing for a greater release of anthocyanin.

Tuberous Vegetables

Fang et al. (2011) evaluated the changes in phenolic compounds during vacuum frying of Chinese purple yam, finding that anthocyanins are sensitive to blanching. Losses of about 60% in anthocyanins were observed, while for phenolics, losses of 30–50% were observed after blanching.

Seeds

Garrido et al. (2008) assessed the phenolic content of almond skins obtained by various processing methods, like blanching combined with freeze-drying, blanching combined with oven drying and roasting. Phenolic content decreased with blanching, yet this effect was less pronounced for blanching/oven drying than for blanching/freeze-drying. Roasted almonds presented the highest phenolic content.

Impact of Blanching on Antioxidant Activity of Foods

Changes taking place in vegetables during cooking may affect their antioxidant activity and free radical scavenging capacity. Such impact may be attributed to heat, processing time and the type of cutting method used (Sultana et al. 2008). In addition, the antioxidant capacity may vary according to the sensitivity of the substance responsible for it to modification or degradation (Bernaert et al. 2014).

Jiménez-Monreal et al. (2009) state that, contrarily to what is observed for the other features of food, antioxidant capacity may increase after cooking. Possible reasons for this behavior include heat-induced synthesis of free radical scavenging antioxidants, release of antioxidant compounds due to cell wall disruption, chemical reactions leading to synthesis of antioxidants and heat-induced inactivation of oxidative enzymes.

Green Leafy Vegetables

In the work of An et al. (2014), they evaluated the antioxidant activity of raw and cooked *Ligularia fischeri*. They found that the higher the blanching time, the lower the product antioxidant activity. They justified their results on the basis of the release of phenolic compounds through ruptured cell walls. Therefore, low blanching times (1–3 min) were recommended for processing the vegetable in order to maintain antioxidant capacity similar to the raw product.

Oboh (2005) reported that green leafy vegetables typically consumed in Nigeria presented an increase in phenolic content and antioxidant capacity after blanching for 5 min.

Cruciferous Vegetables

Minimal heat treatments, like blanching, have been recommended for avoiding high losses of antioxidant properties in cruciferous vegetables (Amin and Lee 2005).

Ahmed and Ali (2013), compared the antioxidant activity of cauliflower by means of DPPH radical scavenging method. Fresh cauliflower presented high antioxidant activity, followed by steam blanched and steam boiled samples. The authors attributed the higher loss of antioxidant capacity during water treatments to the high contact area between cauliflower and water and also to the processing time. It was also evident from this study that the contact area between food and heat is lower during steam and stir frying treatments as compared to water treatments. Leaching may be an important source of loss of antioxidant substances. Cooking also softens and breaks cell components of vegetables leading to release of antioxidant compounds in the cooking water.

Leek

Leek is usually consumed after a heat treatment. Bernaert et al. (2014) elucidated the influence of household treatments, including blanching, cooking and vapor cooking, on the antioxidant properties of white shaft and green leaves of leek. Blanching was performed inside a stainless steel vessel containing 3 l of boiling water and covered with a lid. After 90 s, samples were collected. Boiling treatments were similar to blanching treatments, except for the fact that samples were collected at 10, 20, 40, and 60 min of process. Boiling was performed inside a pressure cooker containing 1 l of water. The authors found that blanching and cooking did not affect the antioxidant capacity of leek white shaft, as confirmed by using the ORAC assay. Blanching of green leaves resulted in a 19% lower antioxidant capacity as compared to raw samples. When leaves were cooked for 40 and 60 min, values of antioxidant capacity (ORAC) were significantly higher than raw samples, namely 12 and 21%, respectively. Such increase may be resultant from the breaking of flavonoids glucosides into aglycones, which present high antioxidant capacity.

With regard to the antioxidant capacity as measured by means of the DPPH method, Bernaert et al. (2014) found that cooking decreased the antioxidant capacity of leek. The authors attributed this behavior to the decomposition of phenolic compounds.

Garlic and Onion

Gorinstein et al. (2008) subjected polish garlic and white and red onions to blanching, boiling, frying and microwaving for various periods of time and assessed the products for bioactive compounds and antioxidant activity. The authors found that blanching and frying of garlic and onion did not affect their bioactive content or their antioxidant capacity.

Fruit

Heras-Ramirez et al. (2009) studied the effect of blanching and drying on features of apple pomace. The drying of both the blanched and the unblanched pomace caused a 60% decrease on the antioxidant activity. However, blanched pomace presented lighter color when compared to the unblanched sample.

Tuberous Vegetables

The effects of blanching, drying and extraction processes on the antioxidant capacity of yam peel were investigated by Chung et al. (2008), who found that blanching at 85 °C for 30 s caused significant decrease in the antioxidant capacity of all of the extracts assessed.

Seeds

The antioxidant capacity of almond skins as obtained by different methods was assessed in the study of Garrido et al. (2008). The use of the ORAC assay showed that the antioxidant capacity was higher for roasted samples, followed by blanched–

dried and blanched–freeze–dried samples. The authors concluded that that roasting was the most suitable way for obtaining almond skin extract of higher antioxidant capacity.

Impact of Blanching on the Vitamin C Content of Foods

Vitamin C (ascorbic acid) is an important component for the human diet which occurs naturally in many foods. It is also used as a food additive. It comprises a substance known for decompose during heat treatment and whose levels can be related to antioxidant activity and shelf life of certain foods (Burdurlu et al. 2006; Bernaert et al. 2014).

Some variability found on the levels of retention of ascorbic acid in food during heat treatment may be attributed to blanching conditions, especially time and temperature, besides brine composition, the vegetable genetic features and maturity. Such factors combined or not, may result in different levels of inactivation of ascorbic acid oxidase enzyme and removal of residual oxygen from the vegetable tissue (Castro et al. 2008).

Green Leafy Vegetables

The ascorbic acid degradation in fluted pumpkin leaves was studied by Ariahu et al. (2011), who found that this phenomenon can be predicted by means of first-order kinetics. Blanching at high pH resulted in higher losses of ascorbic acid. On the other hand, low pH, short time and high temperature produced high retention of ascorbic acid. The authors concluded that blanching in acid medium is a suitable method for processing pumpkin leaves with high ascorbic acid retention.

Peppers

Castro et al. (2008) blanched green and red peppers under various process conditions, finding that the more severe the blanching conditions, the higher the ascorbic acid losses. Losses of 30 and 45% were observed for blanched red and green pepper, respectively.

Tuberous Vegetables

Ascorbic acid is one of the most important nutrients of potatoes. It is used as an indicator of the heat treatment intensity. Blanching usually promotes considerable losses of ascorbic acid (Arroqui et al. 2001).

Several factors contribute to the ascorbic acid loss in potatoes, like enzymatic oxidation, thermal degradation, and diffusion. Nevertheless, diffusion is believed to be the most important loss mechanism of ascorbic acid in potatoes. This process is affected by various factors, such as the presence of components of different molecular mass, solution dilution, and solute gradient (Gekas 1992).

When studying the blanching of potatoes in solutions of different soluble solids content, Arroqui et al. (2001) found that the higher the blanching temperature and time, the higher the vitamin C losses.

Cruciferous Vegetables

Galgano et al. (2007) assessed the effect of household practices like chilling, freezing, and cooking on vitamin C retention in broccoli. They found that blanching at 96 °C for 3 min before frozen storage caused a 32% decrease in vitamin C. They affirmed that steam blanching is more advantageous than hot water blanching since leaching of vitamin C is drastically reduced during the former.

Organosulfur Compounds

Organosulfur compounds are bioactive, health promoting substances that include: three γ-glutamyl peptides, namely, γ-L-glutamyl-S-allyl-L-cysteine, γ-L-glutamyl-S-(trans-1-propenyl)-L-cysteine, and γ-L-glutamyl-S-methyl-L-cysteine; their corresponding S-alk(en)yl-L-cysteine sulfoxides, which are (+)-S-allyl-L-cysteine sulfoxide (alliin), (+)-S-(trans-1-propenyl)-L-cysteine sulfoxide (isoalliin), and (+)-S-methyl-L-cysteine sulfoxide (methiin), respectively; and (1S,3R,5S)-3-carboxy-5-methyl-1,4-thiazane 1-oxide (cyclo-alliin) (Beato et al. 2012).

Garlic

Beato et al. (2012) recommended that, for the processing of garlic pickles, a blanching step is mandatory for inactivating aliinase enzyme responsible for pungent taste and green color in the product. The authors found that blanching at 90 °C for 5 min did not affect significantly the content of individual organosulfur compounds, except from γ-L-glutamyl-S-allyl-L-cysteine e do S-allyl-L-cysteine which slightly decreased. The γ-glutamyl-transpeptidase enzyme is inactivated during blanching, thereby not being responsible for hydrolysis of the evaluated compounds. Even though blanching was directed toward alliinase enzyme inactivation, the γ-glutamyl-transpeptidase enzyme was also inactivated by blanching. Therefore, the contents of γ-glutamyl peptides, S-alk(en)yl-L-cysteine sulfoxides e S-allyl-L-cysteine remain high in garlic pickles.

Leek

Bernaert et al. (2014) showed that the isoalliin content of the leek white shaft blanched for 90 s was significantly higher than that of raw samples. However, after cooking for 10 min, the isoalliin content decrease 41% and after cooking for 40 min all the isoalliin was degraded.

Fructooligosaccharides

Among prebiotic carbohydrates, fructooligosaccharides are associated with therapeutic benefits for diabetes, mineral absorption in the gut and colon cancer prevention (Habib et al. 2011).

Campos et al. (2016) evaluated the stability of fructooligosaccharides and the color of yacon (*Smallanthus sonchifolius*) during blanching and drying. They found that blanching of 5 mm slices during 6 min inactivated polyphenol oxidase and peroxidase activities. Furthermore, blanching in ascorbic acid/CaCl$_2$ solutions prevented reducing sugar and fructooligosaccharides losses and improved the color of the obtained flour.

Final Considerations

Blanching is a commonly used process in the vegetables processing industry. However, vegetables contain expressive amounts of bioactive compounds. The blanching process must be designed in order to inactivate inconvenient enzymes and preserve as much as possible the bioactive compounds contained in vegetables. Fruits are also good sources of bioactive compounds, yet they are not usually blanched. The use of steam blanching is a suggestion for avoid leaching of hydrosoluble bioactives. The use of low temperatures and low times also contributes for preserving thermolabile bioactives. In a well-designed blanching process, the benefit of increase in shelf-life obtained by means of blanching must surpass the losses of bioactive compounds.

References

Adebooye OC, Vijayalakshmi R, Singh V (2008) Peroxidase activity, chlorophylls and antioxidant profile of two leaf vegetables (*Solanum nigrum* L. and *Amaranthus cruentus* L.) under six pretreatment methods before cooking. Int J Food Sci Technol 43:173–178. Doi:10.1111/j. 1365-2621.2006.01420.x

Ahmed FA, Ali RFM (2013) Bioactive compounds and antioxidant activity of fresh and processed white cauliflower. Biomed Res Int. Doi:10.1155/2013/367819

Amin I, Lee WY (2005) Effect of different blanching times on antioxidant properties in selected cruciferous vegetables. J Sci Food Agric 85:2314–2320. Doi:10.1002/jsfa.2261

Amoussa-Hounkpatin W, Mouquet-Rivier C, Kayodé APP et al (2013) Effect of a multi-step preparation of amaranth and palm nut sauces on their carotenoid content and retinol activity equivalent values. Int J Food Sci Technol 48:204–210. Doi:10.1111/j.1365-2621.2012.03178.x

An S, Park HS, Kim GH (2014) Evaluation of the Antioxidant Activity of Cooked Gomchwi (*Ligularia fischeri*) using the myoglobin methods. Prev Nutri Food Sci 19:34–39. Doi:10.3746/pnf.2014.19.1.034

Ariahu CC, Abashi DK, Chinma CE (2011) Kinetics of ascorbic acid loss during hot water blanching of fluted pumpkin (Telfairia occidentalis) leaves. J. Food Sci. Technol. 48:454–459

Arroqui C, Rumsey TR, Lopez A, Virseda P (2001) Effect of different soluble solids in the water on the ascorbic acid losses during water blanching of potato tissue. J Food Eng 47:123–126. Doi:10.1016/S0260-8774(00)00107-2

Beato VM, Higinio A, Castro A, Montan A (2012) Effect of processing and storage time on the contents of organosulfur compounds in pickled blanched garlic

Bernaert N, De Loose M, Van Bockstaele E, Van Droogenbroeck B (2014) Antioxidant changes during domestic food processing of the white shaft and green leaves of leek (*Allium ampeloprasum* var. *porrum*). J Sci Food Agric 94:1168–1174. Doi:10.1002/jsfa.6389

Bernhardt S, Schlich E (2006) Impact of different cooking methods on food quality: Retention of lipophilic vitamins in fresh and frozen vegetables. J Food Eng 77:327–333. Doi:10.1016/j.jfoodeng.2005.06.040

Brambilla A, Lo Scalzo R, Bertolo G, Torreggiani D (2008) Steam-blanched highbush blueberry (*Vaccinium corymbosum* L.) juice: phenolic profile and antioxidant capacity in relation to cultivar selection. J Agric Food Chem 56:2643–2648. Doi:10.1021/jf0731191

Bunea A, Andjelkovic M, Socaciu C et al (2008) Total and individual carotenoids and phenolic acids content in fresh, refrigerated and processed spinach (*Spinacia oleracea* L.). Food Chem 108:649–656. Doi:10.1016/j.foodchem.2007.11.056

Burdurlu HS, Koca N, Karadeniz F (2006) Degradation of vitamin C in citrus juice concentrates during storage. J Food Eng 74:211–216. Doi:10.1016/j.jfoodeng.2005.03.026

Campos D, Aguilar-Galvez A, Pedreschi R (2016) Stability of fructooligosaccharides, sugars and colour of yacon (*Smallanthus sonchifolius*) roots during blanching and drying. Int J Food Sci Technol n/a–n/a. Doi:10.1111/ijfs.13074

Castro SM, Saraiva JA, Lopes-da-Silva JA et al (2008) Effect of thermal blanching and of high pressure treatments on sweet green and red bell pepper fruits (*Capsicum annuum* L.). Food Chem 107:1436–1449. Doi:10.1016/j.foodchem.2007.09.074

Chantaro P, Devahastin S, Chiewchan N (2008) Production of antioxidant high dietary fiber powder from carrot peels. LWT—Food Sci Technol 41:1987–1994. Doi:10.1016/j.lwt.2007.11.013

Chung YC, Chiang BH, Wei JH et al (2008) Effects of blanching, drying and extraction processes on the antioxidant activity of yam (*Dioscorea alata*). Int J Food Sci Technol 43:859–864. Doi:10.1111/j.1365-2621.2007.01528.x

Del Bo C, Riso P, Brambilla A et al (2012) Blanching improves anthocyanin absorption from highbush blueberry (*Vaccinium corymbosum* L.) Purée in healthy human volunteers: a pilot study. J Agric Food Chem 60:9298–9304. Doi:10.1021/jf3021333

Fang Z, Zhang M, Sun Y, Sun J (2006) How to improve bayberry (*Myrica rubra* Sieb. et Zucc.) juice color quality: effect of juice processing on bayberry anthocyanins and polyphenolics. J Agric Food Chem 54:99106. Doi:10.1021/jf051943o

Fang Z, Wu D, Yü D et al (2011) Phenolic compounds in Chinese purple yam and changes during vacuum frying. Food Chem 128:943–948. Doi:10.1016/j.foodchem.2011.03.123

Francisco M, Velasco P, Moreno DA et al (2010) Cooking methods of *Brassica rapa* affect the preservation of glucosinolates, phenolics and vitamin C. Food Res Int 43:1455–1463. Doi:10.1016/j.foodres.2010.04.024

Galgano F, Favati F, Caruso M et al (2007) The influence of processing and preservation on the retention of health-promoting compounds in broccoli. J Food Sci 72:S130–S135. Doi:10.1111/j.1750-3841.2006.00258.x

Garrido I, Monagas M, Gómez-Cordovés C, Bartolomé B (2008) Polyphenols and antioxidant properties of almond skins: influence of industrial processing. J Food Sci 73:C106–C115. Doi:10.1111/j.1750-3841.2007.00637.x

Gekas V (1992) Transport phenomena of foods and biological materials. Boca Raton: CRC Press

Gorinstein S, Leontowicz H, Leontowicz M et al (2008) Comparison of the main bioactive compounds and antioxidant activities in garlic and white and red onions after treatment protocols. J Agric Food Chem 56:4418–4426. Doi:10.1021/jf800038h

Habib NC, Honoré SM, Genta SB, Sánchez SS (2011) Hypolipidemic effect of *Smallanthus sonchifolius* (yacon) roots on diabetic rats: biochemical approach. Chem Biol Interact 194:31–39. Doi:10.1016/j.cbi.2011.08.009

Heras-Ramirez ME, Quintero-Ramos A, Talamás-Abbud R et al (2009) Influence of blanching and drying treatment on total polyphenols and antioxidant activity in apple pomace. In: 5th international technical symposium on food processing, monitoring technology in bioprocesses and food quality management, pp 1007–1015

Jiménez-Monreal AM, García-Diz L, Martínez-Tomé M et al (2009) Influence of cooking methods on antioxidant activity of vegetables. J Food Sci 74:H97–H103. Doi:10.1111/j.1750-3841.2009.01091.x

Maiani G, Castón MJP, Catasta G et al (2009) Carotenoids: actual knowledge on food sources, intakes, stability and bioavailability and their protective role in humans. Mol Nutr Food Res 53 (Suppl 2):S194–S218. Doi:10.1002/mnfr.200800053

Mayer-Miebach E, Spieß WEL (2003) Influence of cold storage and blanching on the carotenoid content of Kintoki carrots. J Food Eng 56:211–213. Doi:10.1016/S0260-8774(02)00253-4

Oboh G (2005) Effect of blanching on the antioxidant properties of some tropical green leafy vegetables. LWT—Food Sci Technol 38:513–517. Doi:10.1016/j.lwt.2004.07.007

Porter Y (2012) Antioxidant properties of green broccoli and purple-sprouting broccoli under different cooking conditions. Biosci Horizons 5:hzs004–hzs004. Doi:10.1093/biohorizons/hzs004

Rodriguez-Amaya DB (2001) A guide to carotenoid analysis in foods. ILSI Press, Washington DC

Rodriguez-Amaya DB, Kimura M (2004) HarvestPlus handbook for carotenoid analysis. IFPRI, Washington DC

Sultana B, Anwar F, Iqbal S (2008) Effect of different cooking methods on the antioxidant activity of some vegetables from Pakistan. Int J Food Sci Technol 43:560–567. Doi:10.1111/j.1365-2621.2006.01504.x

Uenojo M, Maróstica MR Jr, Pastore GM (2007) Carotenóides: propriedades, aplicações e biotransformação para formação de compostos de aroma. Quim Nova 30:616–622. Doi:10.1590/S0100-40422007000300022

Veda S, Platel K, Srinivasan K (2010) Enhanced bioaccessibility of β-carotene from yellow-orange vegetables and green leafy vegetables by domestic heat processing. Int J Food Sci Technol 45:2201–2207. Doi:10.1111/j.1365-2621.2010.02385.x

Zhang D, Hamauzu Y (2004) Phenolics, ascorbic acid, carotenoids and antioxidant activity of broccoli and their changes during conventional and microwave cooking. Food Chem 88:503–509. Doi:10.1016/j.foodchem.2004.01.065

Chapter 4
Use of Blanching to Reduce Antinutrients, Pesticides, and Microorganisms

João Luiz Andreotti Dagostin

Abstract Undesirable substances may occur in vegetable species as endogenous substances called antinutrients. Other undesirable compounds may exist in foods through external sources, due to the contamination by pesticides and microorganisms for example. Some of the positive side effects incurring in blanched foods comprise the removal of these substances and microorganisms, which may imply quality loss, toxicological problems, and infectious diseases. The removal rate of each substance is dependent on parameters related to the process, food characteristics, and substance type. In most processing lines, a few methods are auto sufficient in removing undesirable substances and chemicals from foods, and most of these methods are too aggressive. For microorganisms, sterilization processes can completely remove microbial contamination, but only a fraction of foods can be processed by such a rigorous operation. Blanching is an adequate method that can be used to remove part of the undesirable substances and lower microbial content of raw materials with less changes in their original characteristics. In this chapter, it will be shown and discussed how blanching affects some antinutrients, pesticides, and microorganisms on foods, with a brief presentation of each substance to better understand their role on the physiological disorders caused in humans.

Keywords Blanching · Antinutrients · Pesticides · Microorganisms

Antinutrients

Antinutrients are molecules endogenous to vegetables that play specific biochemical roles in the plant organism, but exert adverse effects when eaten by humans and animals. Some antinutrients are capable of blocking the absorption of nutrients

J.L.A. Dagostin (✉)
Graduate Program in Food Engineering, Chemical Engineering Department, Federal University of Paraná, 19011 Francisco H. dos Santos (S/No), Curitiba 81531-980, Paraná, Brazil
e-mail: joaodagostin@hotmail.com

© Springer International Publishing AG 2017
F. Richter Reis (ed.), *New Perspectives on Food Blanching*,
DOI 10.1007/978-3-319-48665-9_4

when consumed. Others may occur as defensive systems in plants, releasing toxics when the vegetable tissue is damaged. Genetic engineering is used to diminish some antinutritional factors of plant individuals. However, due to their importance in vegetable species, they cannot be completely removed to preserve the plant's own integrity. In this section, blanching will be presented as a method for reducing some antinutrients: proteinase inhibitors, glycoalkaloids, phytates, oxalates, tannins, nitrites, nitrates, and cyanogenic glycosides.

Trypsin and Chymotrypsin Inhibitors

Pancreatic enzymes are a group of substances produced in part of the digestive systems of animals. These enzymes are synthesized by the pancreas (in their inactive forms) and secreted into the duodenum as pancreatic juice (where they become active enzymes) to induce the cleavage of specific substances as fats, proteins, and polysaccharides. Breaking these polymeric molecules is a key biological activity: a proper absorption of nutrients by the body is dependent on the existence of smaller molecules.

Proteases are enzymes capable of breaking the peptide bonds of specific proteins. This specificity of proteins remains in their different structures and amino acids configurations. Two important proteases secreted in the pancreatic juice are trypsin and chymotrypsin, both comprehending serine proteases. Trypsin and chymotrypsin inhibitors are substances that reduce the biological activity of the respective enzymes. Most reports show their occurrence in natural sources such as bovine pancreas and lung, raw egg white, soybeans and lima beans, sweet potato, and peanuts (Vanderjagt et al. 2000; Arcoverde et al. 2014). However, several other plants may also present proteinase inhibitors. The trypsin inhibition may cause not only a lower nutrient absorption, but also play as a cofactor in causing enterocolitis necroticans, a food poisoning generated by a β-toxin of *Clostridium perfringens* type C (Farrant et al. 1996).

Soybean is the main vegetable source investigated in studies where blanching is applied as a method of trypsin and chymotrypsin inhibitors removal. This is due to the higher inhibitory activity presented by soybeans, compared to other legumes and vegetables. The use of leaching and heat treatments in legumes are considered effective technologies not only for the removal of protease inhibitors, but also for inactivating lipoxygenase, an enzyme responsible for producing undesirable beany off-flavors (Shi et al. 2004; Mozzoni et al. 2009).

Using short-time boiling is a good method to reduce trypsin inhibitors in soybean, once several quality attributes can be maintained or slightly reduced. In the work of Mozzoni et al. (2009), for example, a jacketed kettle was used to verify the application of 5 min blanching in soybeans immersed in boiling water. The authors verified that this was enough to achieve a 88% lower trypsin inhibitory activity (resulting in 19 U/g) in either pods or shelled beans. In the mentioned samples, no differences were found for iron, mono and oligosaccharide contents, while higher blanching times (20 min) resulted in reduced values of texture (19–29% lower) and color intensity (10–40% lower).

In another work, Yuan et al. (2008) also verified the use of blanching at different time × temperature sets (0.5–7.5 min; 70–85 °C) in the reduction of trypsin inhibitors activity of soybean milk. The authors found that the application of blanching for 3 min at 80 °C yielded a 49% lower inhibitory activity, compared to the raw soybean. It is relevant to mention that before blanching, the soybeans (Proto variety) were soaked for 15 h at ambient temperature (~ 22 °C), and after blanching, the beans were immediately cooled in water (10–15 °C). The pre-soaking allows the grain to absorb water and improves the conduction of thermal energy. Besides, blanching soybean at 80 °C for 3 min also lowered the hexanal content in the milk to 99% less (0.02 ppm), compared to that of the raw grain (3.48 ppm). Hexanal is a volatile compound that arises from the action of endogenous lipoxygenase and hydroperoxide lyase enzymes over linoleic acid and its products of cleavage, respectively (Hildebrand et al. 1990). The main problem associated with hexanal and some other volatiles in soybean is the peculiar and undesirable beany off-flavor that they confer to food products (Lv et al. 2011).

In 1997, Wang and coworkers conducted a study to verify the combined effects of soaking with water blanching and steam blanching on the nutrients and trypsin inhibitor activity (TIA) of cowpea (Wang et al. 1997). As expected, higher blanching times resulted in lower TIA for both blanching methods. Lower oligosaccharides contents and higher solids loss were found in water-blanched legumes as a result from leaching. Higher soaking times before blanching enhance the reduction of trypsin inhibitors, e.g., when blanching (100 °C) for 20 min, the reduction of TIA was 85% and 72% for trials including an 8 h soaking step and no soaking, respectively. For a 5.87 min blanching combined with a 2.33 min pre-soaking, only 27% less TIA was obtained. With these results we can verify that soaking is a step that helps with trypsin inhibition, but at a very lower extent compared to the effect of blanching itself. Steam blanching had a similar TIA inhibition to water blanching, varying 1–3% on the examples shown above.

Besides the blanching parameters be of great importance in removing proteinase inhibitors, the use of different methods of energy transfer and different raw materials will make blanching behave distinctively in trypsin inhibitors removal. In fact, not only in TIA inhibition, but they will also affect a huge variety of attributes, such as the concentration of nutrients and antinutrients, and sensory aspects. Regarding the trypsin inhibitors presence, plant materials may contain more, less or even not contain the substances (Vanderjagt et al. 2000; Arcoverde et al. 2014). As an example, the total TIA of raw cabbage, collard, turnip, sweet potato leaves, and peanut leaves were 58.5, 60.1, 57.8, 52.0, and 41.0 trypsin inhibitor units per gram, respectively, as found by Mosha and Gaga (1999). Despite receiving the same blanching treatments (water at 98 °C for 2.5–10 min), the authors verified different reductions in trypsin inhibitory activity (Fig. 4.1). Also, the use of a microwave blanching (750 W, 30–60 s) enhanced the removal of trypsin inhibitors, needing much less time to achieve similar TIA inhibition than in conventional blanching. In the same study, the presence of chymotrypsin inhibitors was verified in the target

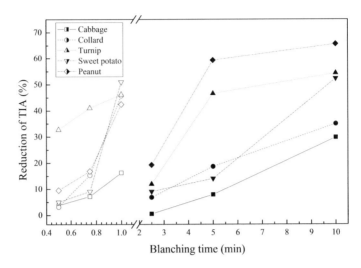

Fig. 4.1 Reduction of trypsin inhibitors activity (TIA) as a function of blanching time. *Open shapes* are microwave-blanched (750 W) trials and *solid shapes* are water-blanched (98 °C) trials. In microwave blanching, samples were placed in water preheated for one minute. All samples represent the leafy parts of the plants [this figure was built by author Dagostin using data from Mosha and Gaga (1999)]

vegetables, and their activity varied from 69.6 chymotrypsin inhibitor units (CIU)/g in peanut leaves to 48.0 CIU/g in collard leaves.

To improve thermal efficiency and water–vegetable contact, Murugkar (2015) investigated the use of a combined grinding-blanching method on the trypsin removal from sprouted soybeans prior to soybean milk and tofu preparation. In this process, the author tested temperature profiles starting from room temperature up to 80, 100, and 121 °C (10, 15, and 25 min to achieve the desired temperatures). After soybean milk and tofu preparation, the reductions of trypsin inhibitors were 62.72, 82.97, and 87.27% for the milk and 73.63, 85.53, and 90.90% for the tofu when processed at 80, 100, and 121 °C, respectively. Despite the fact that the process at 121 °C presented a higher reduction in trypsin inhibitor, the authors considered heating up to 100 °C as an optimal processing since the combined quality parameters presented more favorable results for the final products. Other works where soybean was subjected to high temperature and long time, e.g., 30 min at 80–100 °C, were not considered as blanching, but cooking instead (references omitted). Due to this, they were not considered in this section.

Even for same vegetable species, the amount of nutrients and antinutrients may be variable. While Mosha and Gaga (1999) found both trypsin and chymotrypsin in peanut and sweet potato greens, Almazan and Begum (1996) and Almazan (1995) did not detect these substances in the same type of leafy vegetables. This happens because, for identical cultivars grown in different environments, a great variability of trypsin inhibitors content can be found. Also, it is known that for different parts of the same plant, trypsin inhibitors are not necessarily distributed in an even way

(Liener and Kakade 1969; Bradbury et al. 1985; Almazan and Begum 1996; Mosha and Gaga 1999). Taking this into account, the use of blanching as means of removing proteinase inhibitors is a process that must be used for well-characterized plants and known origin.

Glycoalkaloids

Glycoalkaloids are secondary metabolites that present two definite structures linked to each other: one is composed of a steroid backbone attached to a double carbon ring where nitrogen is usually inserted; and the other is a carbohydrate, attached to the 3-OH position of the steroidal structure (Fig. 4.2). These substances are produced by species of the Solanaceae family to act as a defensive system against diseases caused by fungi, bacteria, and viruses. Their concentration in plants is a function of species and cultivar, geographic distribution, season, cultivation practices, physical injury, maturity, storage temperature, storage under light, and stress factors (Dimenstein et al. 1997; Khan et al. 2013). Glycoalkaloids can be produced in different parts of the same plant, such as in leaves, flowers, roots, and in sprouts (Friedman 2015; Sucha and Tomsik 2016). Some species of Solanaceae are widely used as food sources, including potatoes, tomatoes, eggplant, and peppers in general.

The occurrence of glycoalkaloids in foods may represent sensory and health issues depending upon the concentration in the food source and amount consumed, respectively. Potato tubers, for instance, taste bitter for glycoalkaloids concentrations of 14 mg/100 g or more. In addition to the off-flavor, glycoalkaloid contents higher than 22 mg/100 g also produce a mild to severe burning sensation in mouth and throat (Sinden et al. 1976). Considering the presence of even higher levels, these substances may cause toxic effects on humans and animals. Glycoalkaloids like α-solanine and α-chaconine act as inhibitors of acetylcholinesterase, which affects the regulation of acethylcholine, a neurotransmitter (Roddick et al. 2001). Other injuries caused by glycoalkaloids involve its toxic action on membranes—which leads to cell disruption—and the reduction of intestinal permeability (Patel

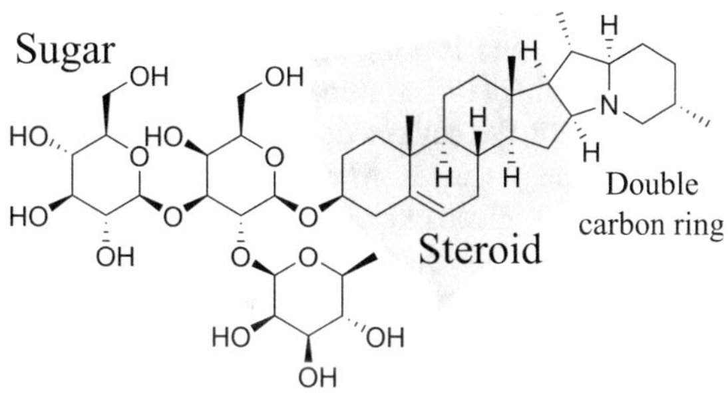

Fig. 4.2 Chemical structure of the glycoalkaloid α-solanine. Solatriose is the sugar fraction, and solanidine comprises the steroidal + double carbon ring fraction (this figure was built by author Dagostin)

Fig. 4.3 Monalisa var.
potatoes exposed to light.
Green parts indicate the
photoinduction of chlorophyll
and glycoalkaloids (this
picture was taken by author
Dagostin)

et al. 2002; Sucha and Tomsik 2016). Common symptoms associated to their
ingestion are abdominal pain, nausea, vomiting, sweating, bronchospasm, halluci-
nation, disorientation, cardiac failure, partial paralysis, convulsions, coma, and even
death may occur (Smith et al. 1996; Friedman 2006).

Potatoes exposed to light after harvesting are stimulated to generate chlorophyll,
which leads to a surface greening coupled with an increase in glycoalkaloid content
by up to three times the original concentration (Dao and Friedman 1994; Phillips
et al. 1996). Figure 4.3 shows Monalisa var. potatoes after the photoinduction of
chlorophyll and glycoalkaloids.

According to Friedman et al. (2003) and Friedman (2015), the majority of
glycoalkaloids of potato tubers are concentrated in the peel, sprouts, and green parts
(when the last two occur), decreasing towards its center. That is why the removal of
their skin helps to diminish in a great extent the total glycoalkaloids content. Also, it
is known that thermal treatments can further reduce their load (Tajner-Czopek et al.
2012). However, the thermal processes must be severe or combined to result in a
significant impact on final glycoalkaloids concentration. In a French fries production line, Rytel et al. (2005) verified the glycoalkaloids (chaconine and solanine)
concentration in two potato varieties (Santana and Innovator) at nine different
processing stages. Non-processed potatoes had a total glycoalkaloid content of 209
and 186 mg kg^{-1} (dry weight) for the Santana and Innovator cultivars, respectively.
With respect to both cultivars, the removal was higher after the stages of peeling
(127 and 121 mg kg^{-1}, 39% and 24% reduction from previous stage), two-stage
blanching (86 and 100 mg kg^{-1}, 36% and 52% reduction from previous stage) and
two-stage frying (14 and 6 mg kg^{-1}, 67% and 83% reduction from previous stage).
Unfortunately, the authors did not describe the parameters of time and temperature
applied in the processes. After the whole process, French fries contained only 3–8%
of the original glycoalkaloids content found in the raw material.

According to the data presented by Lachman et al. (2013), blanching may affect
in distinct ways the removal of chaconine and solanine for different potato cultivars.
The tests were performed in 60–100 g tubers which were peeled (1–2 mm

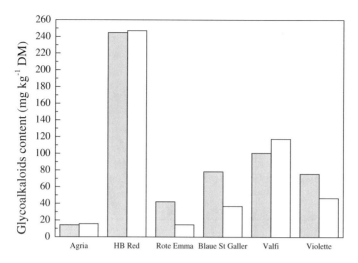

Fig. 4.4 Total steroid glycoalkaloids in fresh peeled (*in gray*) and blanched peeled (*in white*) potatoes of different cultivars [this figure was built by author Dagostin using data from Lachman et al. (2013)]

thickness) and boiled for 15 min. The results of total glycoalkaloids are shown in Fig. 4.4. As can be seen, at least half of the samples achieved low reductions, and others had even higher values of glycoalkaloid content than the unblanched ones. The great variability in the results is probably a consequence of using whole tubers associated with different cultivars. Minimal processing improves the contact area, the leaching rate, and the thermal destruction of substances in water-blanching steps (Rytel 2013).

Even for small pieces of vegetable, blanching cannot be used as unique method in glycoalkaloids removal if a great concentration is meant to be eliminated. The technique is capable of removing part of these substances indeed, but at a limited extent. There are several reports showing that blanching is effective in removing approximately 15–25% of the total glycoalkaloid load of the vegetable (Table 4.1).

All studies found, involving blanching and glycoalkaloids removal, are restricted to potato processing: dehydrated potato in halves (Rytel 2012a), dices (Rytel 2012b; 2013) and granules (Elzbieta 2012), potato flakes (Mäder et al. 2009), French fries (Tajner-Czopek et al. 2012; Rytel et al. 2013; Tajner-Czopek et al. 2014) and potato chips (Pęksa et al. 2006) processing. For steps involving high thermal processes as drying and frying, blanching appears only as an adjunct in glycoalkaloid removal. The highest and really significant amounts of these substances are removed after peeling and contact with more rigorous thermal processes. No study was found to involve the blanching of different vegetables for glycoalkaloids assessment, not even in the use of blanching in unpeeled potatoes prior to producing rustic-type foods. Rustic potatoes are consumed worldwide and, considering the higher glycoalkaloid content present in the peel, it should be a relevant field to be explored.

Table 4.1 Relevant data of blanching studies involving the glycoalkaloids removal from foods

Potato cultivar	Pretreatments	Blanching parameters	Glycoalkaloids removal	Reference
– Denar – Pasat – Innovator (1 × 1 × 1 cm)	– Peeling (1.5 mm); – Cutting; – Rinsing (20 °C)	75 °C/10 min	17.5% (Denar) 12.9% (Pasat) 26.2% (Innovator)	Rytel (2012a)
– Rosalinde – Blue Congo (1 × 1 × 6-7 cm)	– Peeling (1.2 mm); – Cutting; – Rinsing	80 °C/10 min	15.0% (Rosalinde) 28.1% (Blue Congo)	Tajner-Czopek et al. (2012)
– Denar – Pasat – Innovator – Karlena (cut into halves)	– Peeling (1.5 mm); – Cutting	75 °C/15–20 min	27.2% (average)	Rytel (2012b)
n.a. (cut into halves)	– Steam peeling; – Cutting	80 °C/15-20 min	25.6% (average)	Elzbieta (2012)
n.a. (1 × 1 × 1 cm)	– Peeling (1.5 mm); – Cutting	– 60 °C/5 min – 75 °C/5 min – 90 °C/5 min	18.8% (60 °C) 22.2% (75 °C) 33.8% (90 °C)	Rytel (2013)
Karlena (1–1.4 cm slices)	– Steam peeling (220 °C/9 s); – Slicing; – Rinsing	– 60–67 °C/20 min	39.3%	Mäder et al. (2009)
– Herbie 26 – H. Burg. Red – Blue Congo – Vitelotte (1 × 1 × 6-7 cm)	– Peeling (n.a.); – Cutting	75 °C/10 min	14.2% (Herbie 26) 8.6% (H. Burg. Red) 5.4% (Blue Congo) 14.7% (Vitelotte)	Rytel et al. (2013), Tajner-Czopek et al. (2014)
– Karlena – Saturna (1 cm thick slices)	– Peeling (n.a.); – Cutting; – Washing	n.a.	27.0% (Karlena) 23.6% (Saturna)	Pęksa et al. (2006)

n.a. Information or detailing not available

Oxalates, Phytates, and Tannins

Phytates and oxalates are salt forms of phytic and oxalic acids. They are chelating agents capable of binding with metal ions such as calcium, iron, and zinc. Diets rich in oxalates and phytates represent a problem in the bioavailability of such metals. Combined with divalent metals, oxalates form tiny crystals that should be excreted by the urinary tract. However, these crystals may be blocked in the kidney tubules and aggregate, causing the formation of kidney stones (Noonan and Savage 1999). Oxalates may be found in plant tissues as insoluble (predominantly as calcium and magnesium salts) and water-soluble crystals (mainly as sodium and potassium salts) (Siener et al. 2006). Phytates accumulate in plants as they ripen, becoming the major way phosphate and inositol are stored in grains and seeds. Studies indicate that they are capable of forming complexes with proteins, changing their structure, and consequently their solubility, digestibility, and enzymatic activity (Kumar et al. 2010).

Tannins are polyphenols reported as health-promoting agents due to anticarcinogenic, antimutagenic, antioxidant, and antimicrobial properties. Besides all these positive factors they are considered antinutrients since they are powerful iron chelating agents, interfering in its bioavailability (Amarowicz 2007). Tannins are also capable of precipitating molecules as alkaloids and different proteins, interfering with their absorption in the organism (Uusiku et al. 2010).

A great inherent difficulty of assessing the loss of these antinutrients is the variability of their content among vegetable species, parts of the plant, and removal degree. Depending on the antinutrient, regular blanching can generate losses in the order of 40 or 50%—for tannins and oxalates, for example. For phytates the removals are commonly in the order of 20–30% for normal blanching conditions. This low final reduction associated with the high sampling variability (high deviation in many measurements) makes the precise assessment of these antinutrients loss to be a complicated task. As an example of variability between vegetables, the phytates loss of 11 different plants or plant parts ranged from 21.0 to 35.2% as a result of blanching at 100 °C for 1–3 min (Somsub et al. 2008). On the other hand, tannins removal by the same method presented a wider range: 6.9–55.4%.

Different maturation stages can also exert great influence on phytates removal as a consequence of variations in vegetables porosity, composition, and hardness, combined with changes in the diffusivity of molecules. For grass pea var. krab (*Lathyrus sativus*) at earlier stages of maturation, the effectiveness of phytates removal by blanching (96–98 °C for 1–2 min) can be more than twice as effective as that of mature grains. As reported by Lisiewska et al. (2006), unblanched grass peas may present from 420 to 980 mg kg^{-1} of phytic phosphorus from green to mature stages. It is important to notice that the hardness of the peas also doubled after maturation, as a consequence of water loss and tissue modification. This impairs the mobility of (anti)nutrients and may be a cause of lowering from 38.1 to 18.4% the capacity of phytates reduction in green and mature peas by blanching.

Taking into account blanching and even boiling as methods to reduce antinutrients in food, we may consider phytates, tannins, and oxalates as heat-stable compounds (Crossey 1991; Gaugler and Grigsby 2009; Kumar et al. 2010; Muraleedharan and Kripa 2014; Duval and Avérous 2016). The removal of these compounds by thermal treatments is commonly adjusted to simple first-order kinetic models with good fittings, and the effect of temperature over the removal rate may be verified with the aid of the Arrhenius equation, as seen in Chap. 2, Sect. 2.2, Eqs. (2.4) and (2.6) (Crossey 1991; Shi et al. 2009). For thermal treatments involving high temperature and effective degradation, two-step equations and more complex models may be suitable or not from case to case.

A great removal of phytate and oxalate contents from *Moringa oleifera* leaves was achieved after blanching in water at 80–85 °C for 15 min (author called it 'simmering') (Sallau et al. 2012). The concentration of the compounds ranged from 10.3 to 4.4 mg kg^{-1} (57.3%) and 18.4–6.2 mg kg^{-1} (66.3%), respectively, which are considered high reductions. If blanching seems to be an effective way of leaching part of the phytates from leaves, we must look again for blanching processes in other food sources to avoid misleads. An example is the blanching of tree spinach. Tree spinach (*Cnidoscolus acontifolus*) leaves are used as soup condiments and are known for possessing high phytate contents (as 4.79 g kg^{-1}) (Oboh 2005). Blanching in boiling water for 5 min has a little effect over the leaf, contributing to remove only 9% of the total phytate. This may represent a problem due to the great phytate content that remains in the plant. A similar trend was observed in eggplant leaves blanched with the same temperature and time set: the blanched samples contained 8% less phytate than the raw material, with a final 372 mg kg^{-1} (Oboh et al. 2005). Complementarily, the removal of this antinutrient from seeds and beans may be more complicated. In asparagus beans flour for example, blanching for 8 min caused no changes in phytate contents (temperature not reported, assuming boiling), despite the still lower content when compared to tree spinach leaves and the combination of several other subsequent processes necessary to produce the flour, as drying and grinding (Nwosu 2010). Blanching asparagus beans flour did not reduce significant amounts of tannins either.

Leaching is mainly responsible for removing oxalates, phytates, and tannins in a regular blanching step. Higher temperatures in the process will indeed affect positively the diffusion of components. However, the extent of effectiveness is dependent on process type, temperature, vegetable type and its characteristics. One way of improving the leaching of substances involves the increase of food superficial area through methods capable of lowering food size prior to or during thermal processing. The use of a grinding-blanching process at 80 °C (10 min ramp), for example, can reduce the phytic acid content of soymilk from 67 to 51 mg kg^{-1} (24% less) (Murugkar 2015). Blanching itself can cause the rupture of cells and improve the leaching of antinutritional factors, but in a very lower degree than a physical method of particle reduction (Oboh et al. 2005).

Green and white cauliflowers, Brussels sprouts and broccolis presenting oxalate concentrations from 500 to 950 mg kg^{-1} were blanched (5:1, water:vegetable) at 95–98 °C for 3, 3.25, 5, and 3 min, respectively (Korus et al. 2011). After

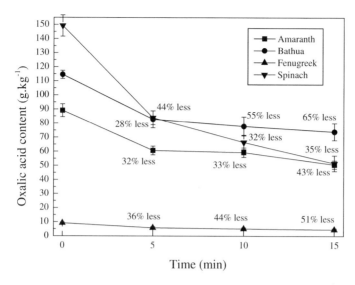

Fig. 4.5 Effect of blanching on the oxalic acid content of leafy vegetables (g kg^{-1}, dry matter). Unknown processing temperature [this figure was built by author Dagostin using data from Yadav and Sehgal (2003)]

processing, true reductions of 6–16% (final 400–860 mg kg^{-1}, fresh weight) of total oxalates contents were achieved. This is a typical set of parameters in a regular blanching, and did not result in a great reduction. From the fractions removed, 23–40% were soluble oxalates. Not only different vegetables present variations on oxalates content, but their content may also vary within different cultivars. Using the same temperature range when blanching Markiza F1 and New Zealand spinach varieties for 2 min resulted in 6% and 30% less total oxalates (to 8.2 and 6.7 g kg^{-1}, d.w.), respectively, while soluble oxalates accounted for 21% (d.w.) of the total for both cultivars (Jaworska 2005). After 5 min processing (unknown temperature, assuming boiling), the total oxalates of regular spinach can be further reduced to values 44% lower (Yadav and Sehgal 2003), as can be seen in Fig. 4.5. Blanching Amaranth, bathua, and fenugreek leaves for this same time can reduce oxalates in 32%, 28%, and 36%, respectively (as oxalic acid). Blanching leaves at higher temperatures and/or for longer can be considered a cooking process due to the substantial changes in the fresh characteristics of the vegetable.

Soluble and insoluble oxalates may also be present in different concentrations for different plants. The reduction of both in blanching processes will also occur in distinct ways as a consequence of their distribution and diffusivity through the plant. In Fig. 4.6 we can see this difference among spinach (Sp), New Zealand spinach (NZ), and silverbeet leaves (Sb) blanched for 2 min in boiling water. The total oxalate removal for the plants was 44.6% (Sb), 25.1% (NZ), and 53.0% (Sp) (Savage et al. 2000). Soluble oxalates are more easily removed from plant tissues by processes involving aqueous solution as a consequence of the facilitated leaching.

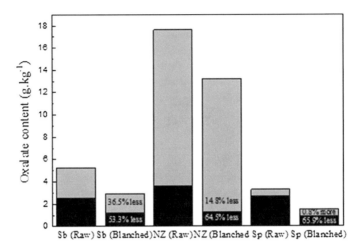

Fig. 4.6 Oxalate content of spinach (Sp), New Zealand spinach (NZ) and silverbeet leaves (Sb) blanched for 2 min in boiling water. Soluble (*black square*) and insoluble (*gray square*) oxalates. Total oxalates are *gray* + *black* bars [this figure was built by author Dagostin using data from Savage et al. (2000)]

A problem associated with leaching of oxalates, phytates, or tannins in processes such as blanching, soaking, or cooking is the concomitant removal of calcium and other minerals like iron, sodium, magnesium, potassium, and phosphorus (Noonan and Savage 1999). In an interesting work, air-dried white *Lasianthera africana* leaves (2 mm width) were blanched in 0.00–1.25% solution of unripe plantain peel ashes (1:3 w/v) at 100 °C for 3 min (Ani et al. 2015). As a result, the tannin contents were greatly reduced as higher the ash concentration in the blanching solution. Tannins contents were 2.4 and 1.4 g kg^{-1} after blanching in pure water and in the 1.25% ash solution, reductions of 14.3% and 50.0% were found, respectively. The authors attributed this effect to a higher leaching of the phyto-chemicals. However, further studies are needed to comprehend this potentiating effect.

The literature still lacks representative works involving the assessment of blanching as a method of reducing oxalates, phytates, and tannins. Some of the studies found were not included in this section due to missing data regarding the blanching parameters, the confusion between blanching and cooking, and erroneous comparisons—as the phytochemical reduction (in fresh weight) between drying processes and blanching, neglecting the water loss. Other works do not evaluate uniquely the blanching process, but the antinutrient removal after two or more processing steps. A great volume of information can be found regarding the loss of antinutrients by cooking, but they can only be considered in parts within the scope of this book. Finally, in this section an analysis of lectins removal was also expected, but it was not performed due to all reasons above-mentioned and the lack of information available regarding blanching.

Nitrite and Nitrate

Nitrites and nitrates are commonly found in processed meat and vegetables in general. For the first food type, they are commonly added as sodium or potassium salts to produce desirable sensorial characteristics and act as antimicrobial agents. In vegetables they naturally occur in varied amounts depending on the plant type, nitrogen fertilizer utilization, season, light, temperature, among other parameters (Chung et al. 2011).

Several risks are associated with the intake of nitrite and nitrate, like methaemoglobinaemia and different cancer types (Jakszyn and Gonzalez 2006; Dubrow et al. 2010; Chan 2011), but most of the studies are still not fully conclusive. After intake, nitrate is converted to nitrite, which can cause the nitrosation of amines, amides, and amino acids—usually found in proteins—, yielding *N*-nitroso compounds such as nitrosamines, nitrosamides, and many others. Due to this, *N*-nitroso compounds can be exogenously found in processed meat (Tricker and Preussmann 1991).

The International Agency for Research on Cancer (IARC) have recently (Oct. 2015) classified processed meat as carcinogenic to humans (Group 1), based on sufficient evidence that the consumption of processed meat causes colorectal cancer (Bouvard et al. 2015). In this news it is cited the formation of *N*-nitroso compounds and polycyclic aromatic hydrocarbons during meat thermal processing as the carcinogenic agents. However, there is still no information about the partial effect of each compound type—the final evaluations will be published in volume 114 of the IARC Monographs. For the intake of vegetables containing nitrates and nitrites, data are still inconclusive about the risks associated. Despite the existence of works presenting the risks of nitrites and nitrates, recent studies also show that nitrate may impose important physiological roles, positively affecting cardiovascular and metabolic functions (Weitzberg and Lundberg 2013). In fruits and vegetables, the formation of *N*-nitroso compounds is inhibited by nitrosation inhibitors, like vitamin C, vitamin E, and polyphenols (Dubrow et al. 2010). However, due to the high variability of nitrate/nitrite and nitrosation inhibitors content of vegetables, the consumption of vegetables with high nitrate and nitrite contents still brings some concerns. Gangolli et al. (1994) reminds that vegetables cannot be treated as if they had the same composition and the same consumption by the population. Additionally, different methods of food treatment may cause different composition not only in nitrate and nitrite content, but also in nitrosation inhibitors. In this section, we will focus on recent studies reporting the nitrate and nitrite loss on food due to blanching.

As an example of the variability of nitrite and nitrate within vegetables, in a study performed by Chung et al. (2011) the nitrite and nitrate contents of 73 different vegetables were $<0.8–9.1$ mg kg^{-1} for nitrite and $<4–6300$ mg kg^{-1} for nitrate in vegetables distributed in different groups (including mushrooms). Those vegetable groups presenting higher nitrate content were leaves, followed by roots and tubers, and finally fruiting and legume vegetables. Three vegetables were

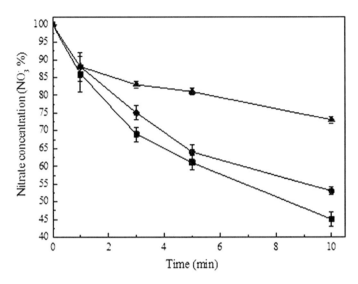

Fig. 4.7 Nitrate concentration (fresh weight) in Chinese flowering cabbage (*black square*), Chinese spinach (*black circle*) and celery (*black up-pointing triangle*) after blanching in boiling water for 1–10 min [this figure was built by author Dagostin using data from Chung et al. 2011)]

chosen for the assessment of the effect of blanching on their nitrate content, namely Chinese flowering cabbage, Chinese spinach and celery (Fig. 4.7). Reductions of 17–31% and 19–39% of nitrates were achieved after 3 and 5 min processing, respectively. However, it must be emphasized that after 10 min the vegetables will probably be cooked and will have lost most (if not all) of their characteristics of freshness. Nitrite levels were in general <1 mg kg^{-1} for the unblanched samples.

Comparable results for two spinach varieties, namely Markiza F1 and New Zealand types, after blanching at 96–98 °C for 2 min were obtained by Jaworska (2005). The difference in nitrate concentration between the two species can be 15-fold or more, for raw or blanched leaves. Nitrate content was reduced by 36% and 23% (to 70 and 1260 mg kg^{-1}, f.w.), respectively to each variety. A very low nitrite reduction (6%) was found for Markiza F1 spinach (to 0.79 mg kg^{-1}, f.w.) while no change occurred in nitrite content of New Zealand spinach.

As small-sized foods improve final nitrate removal, minimal processing prior to blanching should be considered if not affecting the final product. Blanching peeled potato tubers (1.0 cm cubes) for 5 min resulted in 17%, 26%, and 31% less nitrates when temperatures of 60, 75, and 90 °C were used (Fig. 4.8) (Rytel 2013). Complementarily, for the unblanched rinsed cubes there was a small reduction in nitrate content: 142, 130, and 122 mg NaNO$_3$ kg^{-1} for 1.5, 1.0, and 0.5 cm cubes. The effect of potato size on nitrate reduction is evident due to the higher contact area between the outer food layer and the blanching water, blanching solution or steam. For whole potatoes (or cut in half for large size tubers) the reduction of nitrates was only 15% after blanching at 80 °C/15 min (Elzbieta 2012) and 20%

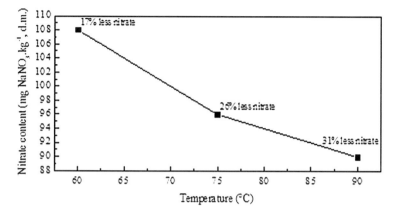

Fig. 4.8 Nitrate content (as sodium salt, dry matter) in potato cubes of 1.0 cm after blanching in water at different temperatures for 5 min [this figure was built by author Dagostin using data from Rytel (2013)]

after blanching at 75 °C/20 min (Rytel 2012a). As for potato dices (1 cm) blanched at 75 °C/5 min, nitrates reductions of 30%, 18%, and 30% were obtained for Denar, Pasat, and Innovator varieties, respectively (Rytel 2012b). The higher nitrate removal of reduced-size potatoes becomes even more prominent if we verify that diced potatoes (1 cm) blanched at 75 °C for only 5 min (Rytel 2012b, 2013) can result in 50–100% higher the nitrate removal than whole/half potatoes blanched at 80 °C for 15 min or at 75 °C for 20 min (Elzbieta 2012). This means that the time needed to process whole tubers is three to fourfold the time used to process diced potatoes to reduce the nitrate content to similar amounts.

Greater changes in nitrate content were observed for cruciferous vegetables blanched at 80 °C for 3 min (Fig. 4.9a). These vegetables—curly kale, green, and white cauliflower, broccoli, and Brussels sprouts—were cut vertically into four or eight pieces prior to immersion, and showed 35–73% less nitrates than the raw samples (Leszczyńska et al. 2009). Compared to the blanching of leaves and potatoes (as shown above), these vegetable types seem to display a higher susceptibility to nitrate loss. As for nitrite reduction a less pronounced effect of blanching was found, and even higher concentrations were verified for Brussels sprouts and curly kale as a result from mass loss after the process (Fig. 4.9b). The levels of nitrites ranged from 1.47 to 3.49 mg kg^{-1} in the raw material, to 1.43–3.28 mg kg^{-1} after blanching, a very little reduction.

In general, significant amounts of nitrates can be removed from plant tissues by blanching. Nitrites cannot be as easily removed. For vegetables this should not be a problem, since nitrites are not present in great quantities. However, the removal of nitrates with adequate technology should be considered for those vegetable with incredibly high nitrate contents, like mustard leaf, spinach, cabbage, and celery varieties.

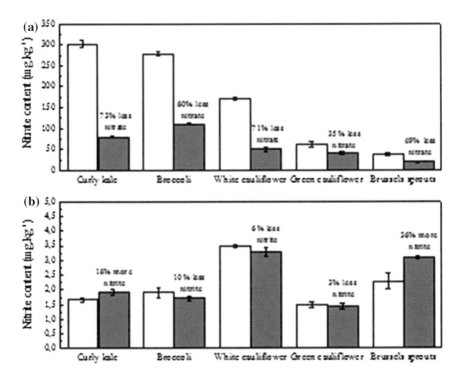

Fig. 4.9 Changes in nitrate and nitrite content (fresh weight) of curly kale, broccoli, white cauliflower, green cauliflower, and Brussels sprouts after blanching in water at 80 °C for 3 min. *White bars* indicate raw vegetables and *gray bars* indicate blanched vegetables [this figure was built by author Dagostin using data from Leszczyńska et al. (2009)]

Cyanogenic Glycosides

Cyanogenic glycosides (CG) are secondary metabolites of plants. As the name says, they are composed of a sugar molecule attached to a cyanide group. They are stored in more than 2500 plant species and act as a defensive system. While the plant is intact, the CG can be stored without causing damage to the cells. If the plant gets hurt or is attacked somehow, cyanogenic glycosides are released and hydrolyzed by enzymes, yielding a sugar and an aglycone. In a second hydrolysis—enzymatic or not—the aglycone releases cyanide, a toxic compound (Vetter 2000).

Long-term intake of cyanide can lead to health disorders like tropical neuropathy, glucose intolerance, an epidemic paralytic disease (konzo), and in some cases can cause cretinism and goiter. The intake of 50–100 mg is characterized as acute poisoning and, although rare, poisoning of this type can be lethal (Montagnac et al. 2009). The direct intake of CGs will also result in the production of cyanide within the organism, so the removal of cyanide-containing molecules from vegetables must be considered when present in large quantities.

Blanching can act in CGs removal as a process of leaching and can further optimize the removal of cyanides through the volatilization of the cyanide compounds. The hydrogen cyanide (HCN) is volatile at temperatures near 25 °C and can be easily evaporated using thermal processes. However, the use of high temperatures can denature the enzymes involved in CG hydrolysis, which can hinder the removal of cyanogenic molecules.

Bitter apricot (*Prunus armeniaca*) seeds for example contain about 50–150 μmol CG per gram (d.w.), composed mainly by amygdalin and prunasin glycosides. Tunçel et al. (1998) verified that the use of blanching in raw seeds containing 39.5 μmol CG/g could reduce in 22–34% the content of cyanide equivalents, probably by leaching (Fig. 4.10). After blanching the seeds were soaked (25 °C for 2 h) and ground to be assessed for the enzymatic activity. There was a slight enzymatic activity reduction after 5 min of blanching, a great reduction after 10 min and no activity after 20 min of thermal treatment, as a result of enzyme denaturation. The effect of blanching per se happened in the first 5 min (33% reduction), making longer processing unnecessary. It is interesting for the enzymes not to be denatured in this case—contrarily to regular blanching processes—so they can act in the release of HCN, which can be more easily removed later.

Removing cyanide compounds from seeds may be difficult due to the higher resistance involved in diffusion of molecules. In the case of blanching asparagus bean (unknown temperature) prior to flour production, no significant differences in

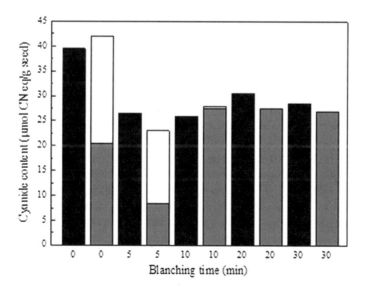

Fig. 4.10 Effect of blanching bitter apricot seeds at 100 °C for different times. Glycoside cyanins right after blanching (*black square*); glycoside cyanins after blanching + soaking (25 °C for 2 h) + grinding (*gray square*); and non-glycosidic cyanide after blanching + soaking (25 °C for 2 h) + grinding (*white square*) [this figure was built by author Dagsotin using data from Tunçel et al. (1998)]

cyanide concentration were found between the finished product and the raw material (Nwosu 2010). According to the author, reductions of cyanide were only 1.9% and 4.8% after 4 and 8 min of blanching.

For most of the cases where leaves are blanched, CGs are removed with a higher efficiency than they are from seeds. As an example, blanching (80–85 °C for 15 min) can remove a great parcel of cyanide from *M. oleifera* leaves; despite the fact that they do not possess a high content of CGs. As reported by Sallau et al. (2012), reductions from 4.2 mg kg^{-1} to 0.8 mg kg^{-1} can be achieved in the process (80.1% efficiency). Leaves of eggplants also contain small amounts of cyanogenic glycosides (Oboh et al. 2005). Raw leaves with 2.0 mg kg^{-1} lost 65% (0.7 mg kg^{-1}) of the cyanide content after blanching in boiling water for 5 min.

Vegetables of different species may have distinct losses of cyanide in blanching processes. Nkafamiya et al. (2010) verified that the cyanide content of 14 leafy vegetables found in Nigeria were from 0.35% to 3.05%. After blanching (unknown parameters) these vegetables, 6.25% to 78.9% of the total amount of cyanide could be removed. This variability occurred not only due to the difference among vegetables but also due to variations in same vegetable types, verified by a great deviation over the mean results found.

Cassava roots are a rich source of carbohydrates (80–90%). In turn, their leaves have a good amino acid balance, are rich in vitamins B1, B2, C, and carotenoids, and are important sources of iron, zinc, manganese, magnesium, and calcium. However, cassava is a vegetable that may contain high levels of cyanide. In the roots—most consumed part of the plant—the cyanide content may range from 10 to 500 mg HCN equivalents kg^{-1} (d.w.), and for leaves this concentration may achieve 53–1300 mg HCN equivalents kg^{-1} (d.w.) (Montagnac et al. 2009). Blanching (boiling water, 5 min) the varieties oko iyawo, IITA white (TMS 30555), and IITA red (TMS 30572) of cassava leaves had no effect on their final HCN potential. Despite the high content of cyanide and the high consumption of cassava roots in tropical and subtropical countries, there is little or no information regarding the effect of blanching on CG or cyanide removal. However, data regarding cooking processes can be found. According to Nambisan and Sundaresan (1985), the process of cooking (boiling water, 30 min) can be more effective than baking (110 °C for 20 min), frying (unknown temperature, 5 min), steaming (pressure cooker, unknown temperature, 20 min), and drying (70 °C for 24 h) on reducing cyanoglucosides from cassava tubers. Three cultivars were assessed by the authors and all presented similar reductions in CGs, with mean reductions of 46.4%, 14.2%, 11.3%, 15.7%, and 27.2%, respectively, to the methods above-mentioned. Also, in the same work the authors report that using larger water volumes in the process of cooking results in a higher CG removal, reaching 77.7% for a 1:10 (root:water) ratio. Complementarily, minimum processing improves the removal of cyanide compounds: the smaller the tuber fractions, the higher the reductions after boiling: 25.0%–74.4% for 50–2 g pieces. Analyzing the process of boiling, it is likely that blanching could also be a feasible alternative in the removal of cyanide and its precursors from cassava roots. It is interesting to note that a common method used in kitchens and restaurants to prepare the root is the application of pressure cooking, which seems to remove just a little of total cyanide content. For the case of a satisfactory removal, blanching could be considered as a previous step of CG removal, prior to the regular pressure cooking. As the effect of blanching on cassava roots is so far unknown, that is just an assumption.

Pesticides

Pesticides are chemical or biological substances intended to kill, incapacitate, or at least keep away a specific group of plagues that infest crops. Infestations reduce food production, affect negatively their overall quality, and may impose harm to human and animal health through disease spreading. Some of the main pesticide types include herbicides, insecticides, and fungicides: herbicides are used to kill weeds and unwanted plant types; insecticides have the purpose of killing insects and arthropods; and fungicides are meant to kill fungi (Tadeo et al. 2008). Several other types of pesticides may be found for different pests.

The great majority of pesticides are of chemical nature. To improve efficacy, they are usually designed to be retained on the vegetable surface. As the pesticide is maintained intact on/in the vegetable, the longer the preventive action will last. This is a very wanted characteristic from the economic and operational point of view, but may become a toxicological problem for the final consumer.

Pesticides that remain on or in food after harvesting are denominated residues. The concentration of these residues in food and foodstuffs are regulated by different agencies all around the world, since they represent health issues. The exposure and bioaccumulation of pesticides can occur by long or short-term high-level exposure —as in the case of farmers and manufacturers in direct contact with concentrated substances—and long-term low-level exposure—which occurs when in contact with material of lower pesticide concentration, as soil, water, and food. The human exposure to pesticides occurs mainly through the diet (Hill et al. 1995). One way of reducing pesticide in food is using technological methods capable of removing pesticides or degrading/breaking the chemicals into non-harmful substances.

Pesticides may be removed by processes involving contact with water depending on their solubility, usually given by the octanol/water partition coefficient (K_{ow}). This coefficient estimates the lipophilicity of certain component through the measure of its affinity to octanol with respect to water. It is not only considered as an index of how a component can be solubilized to water, but also as a predictor of its ability to pass through biological membranes and to accumulate in living tissues (Vighi and Guardo 1995). The K_{ow} can be calculated by

$$K_{ow} = \frac{P_i^o}{P_i^w} = \frac{P^o x_i^o}{P^w x_i^w} \tag{4.1}$$

where P_i^o and P_i^w are the molar concentrations of the i component (pesticide, in this case) in the 1-octanol-rich and water-rich phases, respectively; x_i^o and x_i^w are the mole fraction of solute i in both phases, respectively; and P^{o} and P^{w} are the total molar concentration of components. In the logarithmic form, Eq. 4.1 becomes

$$\log K_{ow} = \log P_i^{o} - \log P_i^{w} \tag{4.2}$$

The higher the log K_{ow} values, the higher the hydrophobicity of a target substance and more difficult should be its removal with water. Vighi and Guardo (1995) suggest a classification scheme where for

Bioaccumulating substances	$\log K_{ow} > 3.5$
Low bioaccumulation potential	$3 < \log K_{ow} < 3.5$
Nonbioaccumulating substances	$\log K_{ow} < 3$

taking into account that this is only an indicative of bioaccumulation potential of a substance, since living tissues are complex systems that depend upon several biological factors. For pesticide removal this trend may be representative to some extent, but cannot be used as the only form of estimation: the removal will be dependent on solubility, food characteristics, process parameters, pesticide and adjuncts characteristics, and many other factors. The effect sizes of log K_{ow} in removing pesticides may be different from those involving bioaccumulation though. In foods where a systemic pesticide was applied, for example—those capable of being absorbed into the vegetable tissue—the removal of pesticide by thermal processes can be compromised if the vegetables are kept whole, but it can be improved for smaller food particles (Kwon et al. 2015).

Some parameters for the log K_{ow} equation have been proposed for estimating the coefficient values for mixed classes of pesticides, but for other chemical types the results may serve only as preliminary screening of the compound removal capability (Finizio et al. 1997). When used for specific compounds, the accuracy of K_{ow} values must be higher. In a study involving the removal of eight different pesticides from cowpea, Huan et al. (2015) verified that for the chemicals presenting log K_{ow} of 2.9–4.4 the use of blanching (100 °C, 200 g L^{-1}) was totally effective after 5 min of processing (Fig. 4.11). For the remaining pesticides, which presented log K_{ow} of 6.0–6.9, the concentration actually raised 4–13% in the final samples. Using a simple washing (25 °C) did not result in the effective removal of pesticides. After 5 min washing, only 6 and 31% of procymidone and chlorothalonil were removed, respectively. According to these results, water blanching may be a useful method to remove pesticides presenting favorable solubility in aqueous media.

In the same work above-mentioned, the use of frying/stir-frying revealed an inverse trend of that from Fig. 4.11: higher K_{ow} pesticides were completely removed from cowpeas after 50 s/3 min cooking, respectively, while for low K_{ow} pesticides no removal was detected. Thus, it is suggested that processes combining blanching + frying/stir-frying may completely remove the pesticides evaluated. The lipophilic (high K_{ow}) molecules were found to be present in the frying oil, and it is very likely that the removal of hydrophilic molecules in blanching also occurs by leaching. However, it is uncertain if the pesticides were completely removed or

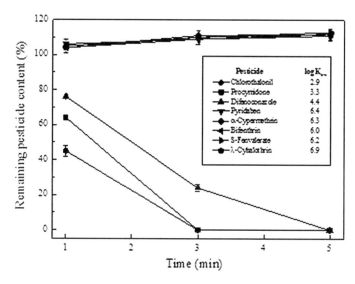

Fig. 4.11 Final concentration of eight pesticides in cowpea after blanching at 100 °C. Initial pesticide concentration of 0.4–0.6 mg/kg [this figure was built by author Dagostin using data from Huan et al. (2015)]

transformed in other chemical compounds. Further investigations must be done to know the existence of breakdown products, which may be even more toxic than the original pesticide substance in some cases.

Blanching was also proven to be a very efficient method for removing profenofos [insecticide, log K_{ow} = 4.44 (Tadeo et al. 2008)] from sweet green pepper and eggplant (Radwan et al. 2005). Initial insecticide content was 0.31 and 0.17 mg kg^{-1} for pepper and eggplant fruits, respectively. After water blanching (boiling for 5 min) the levels of profenofos decreased to 0.006 mg kg^{-1} (98% less) in sweet green peppers and to non-detectable amounts in eggplants.

In 2010 a meta-analysis statistical method was applied to calculate the combined effect sizes of different food processes on the pesticide removal of several vegetables (Keikotlhaile et al. 2010). The analysis was performed using the data of 33 works judged as relevant. According to the authors, blanching, cooking, frying, peeling, and washing are processes that have great influence on the reduction of pesticide residue levels of foods. While peeling removes the residues from the outer layer of vegetables (usually containing higher concentrations of pesticides) washing, blanching, and other thermal treatments are capable of leaching and degrading pesticides depending upon their chemical type and treatment parameters (Bajwa and Sandhu 2014).

Every step of a processing line involving the washing, blanching, and microwave cooking or in-pack sterilization of spinach leaves (*Spinacia oleracea*) was capable of reducing the concentration of at least four of the following pesticides: boscalid [log K_{ow} = 2.96, van Ravenzwaay and Leibold (2004)], mancozeb (log

K_{ow} = 1.33, Toxicoloy database of the U.S. National Library of Medicine), ipro-dione [log K_{ow} = 3.00, Gómez et al. (2011)] and propamocarb [log K_{ow} = 1.12, Chamberlain et al. (1996)] (fungicides) and deltamethrin [log K_{ow} = 4.60, Kaneko and Miyamoto (2001)] (insecticide), according to Bonnechère et al. (2012). The step of washing (15 °C for 3 min) reduced pesticide residues by 10–50%, while blanching (88 °C for 5 min) allowed reductions up to 70%. Blanching was the step of higher effectiveness in the removal of pesticide residues from spinach. From the chemicals tested, deltamethrin (highest K_{ow}) was neither removed by washing nor by blanching; its content was even greater after processing. The higher pesticide removal in the blanching process was achieved in samples contaminated with propamocarb, followed by iprodione, boscalid, and mancozeb. The authors have found an increased 3,5-dichloroaniline content in the samples after all processes. This substance is a toxic product of degradation from iprodione, and was increased from 0.046 mg kg^{-1} to 0.076 mg kg^{-1} after washing and blanching, which is still a low concentration. Ethylene thiourea, a product from mancozeb degradation, was also monitored but was not detected in spinach samples.

Different water volumes [4:10, 3:10, 2.7:10, and 2.5:10, plant material (mL): water (mL)] were verified to have little or no influence on the reduction of pyrid-aben concentrations in hot pepper fruit and leaves by the use of blanching at 100 °C for 1 min (Kim et al. 2014). The application of pesticides in single dose rates (1000 times dilution) made the fruit present 70.1–75.1% (final 0.91–0.87 mg kg^{-1}) less pesticide than the unprocessed fruits. Leaves presented even higher reductions: 85.1–90.5% less pyridaben after blanching. Using different holding times (1–4 min) of immersion also did not result in greater pesticide removal. These results differ from those found by Huan et al. (2015)—where pyridaben was not removed from cowpeas after very similar blanching conditions (Fig. 4.11). In this case, a great part of pyridaben was removed from the fruits and leaves of hot pepper. Regarding only the information available, a possible explanation to this difference is that pyridaben can attach to the surface of vegetables in distinct ways or can be removed more easily depending on the vegetable type.

Microorganisms

Convenient or processed foods are those semi or fully prepared foods that emerged to attend consumers seeking ease in preparing meals. As nutritive media, these foods can be contaminated with biological organisms at different steps of the processing chain, such as harvesting, transporting, cleaning, peeling, cutting, mixing, and handling in general. Even processes expected to prevent contamination are susceptible to raise the microbial load of foods, like washing with contaminated water and using inappropriate or contaminated packaging, seals and equipment in general. For processed foods, a minimum quality regarding their safety and healthy, nutritive and sensorial characteristics is not only expected, but demanded. Spoilage caused by microorganisms is one of the main factors affecting all characteristics above, but concerns primarily due to the potential of causing foodborne diseases. In order to diminish the increase of microbial contamination, in-industry quality may be improved by the application of good manufacturing practices, the application of quality programs (e.g., HACCP and pest control) and standard operations of manufacturing, cleaning, and sanitizing, for example. But even taking the necessary care, it is impossible to reduce microbial counts in foods if no processing is applied. Some industrial processes are intended to exclusively reduce microbial population, as the application of ionizing radiation, the use of specific additives, sterilization, and in some cases microfiltration and modified atmosphere packaging. However, most of the processes used in the food processing plants have more than one purpose or affect more than one characteristic of the product.

Among all processes types, those involving the exposure of foods to mid to high temperatures are preferred for microbial destruction. Cooking, roasting, frying, sterilization, pasteurization, tyndallization, drying, evaporation, blanching, extrusion, and hot extraction are examples of thermal processes. These methods can be controlled to provide multiple special characteristics of texture, nutritional content, flavor, color, and microbiological safety to processed foods.

Blanching is a great example of multipurpose industrial method. Its main function is the destruction of endogenous enzymes that may impair sensorial characteristics, but it can also be a method of peeling, a method of removing undesirable substances and also microbial loads. Due to the time and temperature of regular blanching, it cannot achieve sterilization, but can reduce a considerable amount of undesirable microorganisms in food material. The application of blanching will affect in some degree the viability of vegetative cells of most microorganisms, including pathogens. Spores, some thermophiles, and hyperthermophiles may survive regular blanching processes.

The sum or synergistic effect of different processes is always an alternative to further decrease microbial population. In a practical example, the dehydration of figs in a convection oven at 40, 60, and 70 °C resulted in yeasts and mesophilic aerobic bacteria counts of approximately 3.0, 2.5, and 2.0 log CFU g^{-1} and 5.0, 4.2, and 3.8 log CFU g^{-1}, respectively to microbial types and temperatures. When blanching (100 °C for 1 min) was applied subsequently to this process, bacteria and yeast counts decreased below the detection limits (Villalobos et al. 2016). Blanching in this case was much more effective against microorganisms than drying, even for a very short time (drying took hours, but precise information is lacking). In the same work, molds formed approximately 3.8 and 3.0 log CFU g^{-1} in figs dried at 40 and 60 °C, but blanching was less pronounced in removing the microbial load compared to the other microbial types, accounting for a reduction of ~ 0.8 and 0.5 log CFU g^{-1}, respectively. After drying at 70 °C, the use of blanching reduced the mold population from 3.5 log CFU g^{-1} to levels below the limit of detection. This greater effect could be a consequence of the higher pre-injury caused by the drying process.

Combinations of ultrasound and blanching can also be performed to improve microbial destruction without extending the changes caused by over blanching. Blanching (90 °C for 5 min) of bottle gourd (*Lagenaria siceraria* var. Kashi

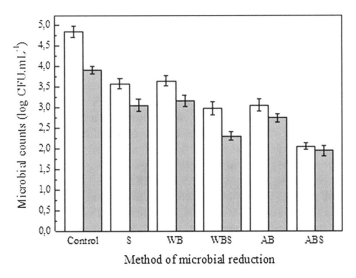

Fig. 4.12 Effects of blanching and sonication on total microbial counts (*white*) and yeast and mold counts (*gray*) in carrot juice. Acronyms indicate sonicated samples (*S*), water blanched samples (*WB*), water blanched and sonicated samples (*WBS*), acid blanched samples (*AB*), and acid blanched and sonicated samples (*ABS*). Blanching at 100 °C/4 min in water (*WB* and *WBS*) or in 45 g/L citric acid solution (*AB* and *ABS*), sonication at 15 °C/2 min with 5 s on and 5 s off intervals at 20 kHz and 70% amplitude level (*S, WBS* and *ABS*). Data are means ± SD (*n* = 4) [this figure was built by the author Dagostin using data from Jabbar et al. (2014)]

Ganga) cubes (2 cm^3) prior to juice extraction was effective in reducing from 1.5 to 0.9 log CFU mL^{-1} the total viable cell count of the product (Bhat and Sharma 2016). The application of a posterior sonication process (50 kHz, 70% amplitude, 25 °C, and 20 min) changed the total microbial count of bottle gourd juice to undetectable levels. For carrot juice production, the use of sonication after blanching was also efficient in microbial reduction (Fig. 4.12) (Jabbar et al. 2014). The use of an acidic blanching solution with blanching improved the reduction on microbial counts. Combining the acid blanching with sonication had an even higher destructive effect over microorganisms. The interesting thing in using sonication with immersive blanching is that the processes can be easily combined: both make use of a water (or solution) bath to transfer thermal energy (blanching) and generate cavitation (ultrasound). For processes of low to mid thermal transfer (as blanching), using a combination of barriers to microorganisms is a feasible alternative. Acid blanching can be either a good alternative to be used in foods meant to be transformed into juices, extracts or purees in general—products where some acidity may be considered a desirable characteristic. In naturally acidic products the use of acid blanching can also have minimum impact on sensorial characteristics.

Whole products were also tested for microbial reduction using a simultaneous blanching + sonication (thermosonication) process (Alexandre et al. 2011). Three different microorganism/food combinations were used to assess the microbial

Fig. 4.13 Effects of regular blanching (*white*), thermosonication (*gray*) and UV-C radiation (*hatched*) on **(a)** *Listeria innocua*/red bell peppers, **(b)** total coliforms/watercress and **(c)** total mesophiles/strawberries. N_0 is the initial microbial concentration and N is the microbial concentration after 2 min of treatment. Data are means $\pm$ SD ($n = 4$) (Alexandre et al. 2011)

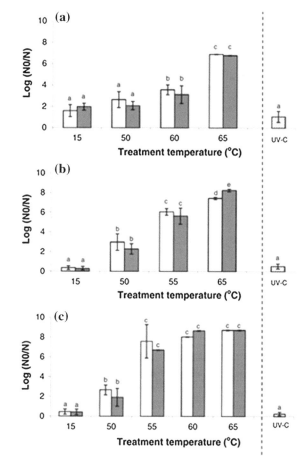

reduction: (a) *Listeria innocua*/red bell peppers, (b) total coliforms/watercress, and (c) total mesophiles/strawberries (Fig. 4.13). The processes applied were regular water blanching (50–65 °C for 2 min), thermosonication (50–65 °C for 2 min, 35 kHz and 120 W), and UV-C radiation (12.36 W m^{-2} for 2 min). In this particular case, no significant differences between blanching and thermosonication occurred. This is probably due to the short time of the process, which is effective for the thermal effect but not enough for the sonication to be significant. However, it is evident the importance of blanching temperature on microbial reduction. The UV-C radiation was the least effective method, achieving reductions of 0.26, 0.53 and 1.05 log CFU g^{-1} in mesophiles, *L. innocua* and total coliforms, respectively—amounts comparable to results obtained by simple washing at 15 °C. While the ultrasound and UV-C methods show higher effectiveness when applied for several minutes (Gómez et al. 2010), blanching can severely reduce microbial loads if tuned for mid-to-high temperatures with short time sets.

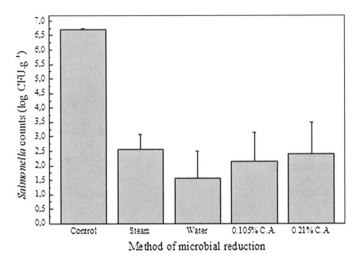

Fig. 4.14 Salmonella counts in carrot slices after blanching by different methods: control (non-blanched samples), steam (steam-blanched samples, 88 °C for 10 min), water (water-blanched samples, 88 °C for 4 min), 0.105% and 0.21% C.A. (samples blanched in citric acid solution, 88 °C for 4 min). Samples were plated in xylose lysine deoxycholate agar [this figure was built by author Dagostin using data from DiPersio et al. (2007)]

Depending on the degree of processing, i.e., the level of each parameter, different blanching methods may or may not present an expected effect on microbial reduction. Figure 4.14 illustrates the effect of different blanching methods on microbial counts of carrot slices (5 mm, crosswise) contaminated with a pool of five *Salmonella* strains—four *S. typhimurium* strains and a *S. enterica* serovar Agona strain (DiPersio et al. 2007). The methods used involved conventional water blanching (88 °C for 4 min), acid blanching (0.105% and 0.21% citric acid solutions, 88 °C for 4 min) and steam blanching (88 °C for 10 min). Mean reductions of ∼4 log CFU g^{-1} where achieved, but no significant differences in *Salmonella* spp. counts were observed among any of the different methods applied. *Salmonella* strains as *S. typhimurium* can be acid-tolerants (Foster 1993), a reason why the acid blanching did not have an additional effect. Despite the use of a very superior processing time in steam blanching, the low temperature may have influenced the effectiveness of the microbial reduction. Steam blanching is known for the higher heat transfer when saturated steam is applied, i.e., for any process conditioned to temperature and pressure ranges where water and vapor can coexist. For these processes, latent heat is transferred when the vapor is condensed on the food surface, markedly improving thermal transference. In this case, steam blanching at 88 °C may not be sufficient to improve efficiency. Taking into account absolute values, regular water blanching was the most effective blanching method in reducing *Salmonella* cells.

Salts are common and cheap microbial inhibitors regularly used in food. The use of some calcium salts in the blanching water may also improve the destruction rate of *Escherichia coli* O157:H7. On spinach leaves, the use of 1% calcium ascorbate or Ca(OH)$_2$ in a soaking solution (22 °C for 5 min) can produce reductions on *E. coli* counts from ~7 log to ~0.8 log CFU g^{-1} or to undetectable values, respectively (Kim et al. 2015). The 1% calcium carbonate, calcium chloride, calcium citrate, calcium gluconate, calcium lactate, and calcium propionate salts solutions did not exhibit bactericide effect against *E. coli* O157:H7 at 22 °C. When using calcium ascorbate and calcium hydroxide solutions at higher temperatures (blanching), the interval needed to inactivate most of the cells was reduced to periods lower than 1 min, which makes this an interesting process for time saving and overcooking reduction. Figure 4.15 shows this bactericidal effect. The use of calcium hydroxide substantially potentiates the destruction of *E. coli* O157:H7 at 45–70 °C. Combining different times, temperatures, and Ca(OH)$_2$ concentrations, Kim et al. (2015) also adjusted a quadratic model to predict the effect of this blanching process (model not shown, R^2 = 0.98), and verified that an optimum blanching condition should be 61.9 °C with 0.52% Ca(OH)$_2$ for 41.7 s, where 5.41 log CFU g^{-1} of *E. coli* O157:H7 may be removed from fresh-cut spinach with a very low weight loss (1.29%). However, other quality parameters should be further investigated to understand sensory variations caused by salt addition.

The use of a sequential blanching may also be an alternative to further reduce microbial counts. In a potato processing (Russet Burbank, 11 × 11 × 50 mm strips), Oner and Walker (2011) performed a two-step blanching combined to a near-aseptic packaging as an attempt to reduce the final microbial counts of the samples. For the first blanching step, 60 °C for 20 min were applied in a CaCl$_2$ solution (0.5 g.L^{-1}). In the second step, blanching occurred at 98 °C for 1 and 5 min. After 28 days, mesophilic bacteria, psychrotrophic bacteria and yeast-mold

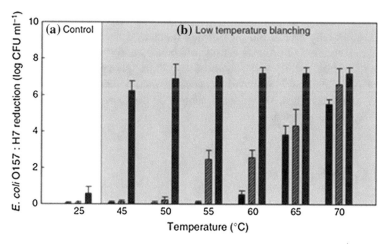

Fig. 4.15 Effect of blanching with 1% calcium ascorbate or 1% calcium hydroxide solutions on the viability of *E. coli* O157:H7 cells in fresh-cut spinach leaves, at different temperatures for 1 min. (*black square*) pure water, (*square with lower left to upper right fill*) 1% calcium ascorbate, and (*square with diagonal crosshatch fill*) 1% calcium hydroxide (Kim et al. 2015)

counts were lowered from counts higher than 5.5 log CFU g^{-1} (for 1 min in second blanching) to values lower than the minimum detection level (<1.95 log CFU g^{-1}, for 5 min in the second-step blanching). For foods of bigger size, the use of a two-step blanching may be considered a multipurpose method, especially when two different temperature/time combinations are applied. While a low-temperature and long-time process (LTLT) may be necessary to remove, destroy, and leach unwanted substances and enzymes from the inner and outer layer of foods, the high-temperature and a short-time process (HTST) comprises a more effective parameter set in destructing superficial microbial contamination. An example of HTST blanching is reported in the work of Pao et al. (2008), where naturally occurring yeasts and molds (initial 3.4 log CFU g^{-1}) and coliforms (initial 2.3 log CFU g^{-1}) of "edamame" (in-pod immature soybeans) were completely removed after blanching at 98 °C for 30 s. Blanching for 60 s removed ~ 6 log CFU g^{-1} of *E. coli* and *Listeria* strains inoculated on fresh edamame, a high reduction for a short period.

When one thinks in blanching, it is almost unconscious to imagine a vegetable being processed. But in the edible insects processing or insects-derived ingredients production, blanching is a feasible alternative for reducing microbial contamination, along with chilling and drying (Marshall et al. 2016). The blanching of yellow mealworm larvae (*Tenebrio molitor*), for example, can cause reductions of 4.4–6.4 log cycles in total aerobic microorganisms for processing periods of 10, 20, and 40 s (for initial counts of ~ 7.9 log CFU g^{-1}) (Vandeweyer et al. 2017). As for *Enterobacteriaceae*, lactic acid bacteria and psychrotrophic aerobic counts, an even greater reduction was achieved, from a mean count of 7.0 log CFU g^{-1} to values in the range of <1.0–1.5 log CFU g^{-1}. The lower reduction was verified for yeasts and molds, with a mean reduction of 1.6 log CFU g^{-1}. Insects have a relatively high superficial area and possess high protein content. Proteins may denature under common blanching temperatures (e.g., 70 °C), so care should be taken when blanching not only insects but protein-rich foods in general, as beef and fish. The thermal denaturation of proteins causes pronounced and irreversible changes in the structure, texture, color, and flavor of the food, giving to it a clear cooked appearance. Blanching is intended to cause minimum change to the fresh characteristics of foods. Thus, the process cannot be confused with cooking itself, a commonly found mistake.

Final Considerations

Regular consumers of processed foods just do not know most of the undesirable substances shown in this chapter. These substances and their associated problems are not as publicized as those of trans-fat, cholesterol, sodium, sugar, artificial additives, and sweeteners, for example. It is not common to hear that a vegetable may be harmful to someone's health. However, the fact that some non-processed foods—including vegetables—may impose risks cannot be ignored by food processing units, who are directly responsible for what living beings are consuming. Furthermore, more information should be given to consumers with respect to the problems associated with certain vegetable types and to the household methods that can be applied to solve these problems. For someone seeking the removal of some undesirable components or lower the microbial load of raw foods, the use of blanching can be a fast and easy alternative to get a safer product and extend its shelf life.

Knowing your raw material is the first step before setting up a blanching process. Enzymatic, sensory, microbial, nutritional, and antinutritional aspects must be evaluated during the process for each raw material, taking into account its size and blanching parameters. Higher processing should only be applied for those vegetables known by having considerable amounts of undesirable substances or microbial loads, in order to avoid over blanching, save time, energy, labor work, and of course, save money. For pesticides it is also important to know the chemical compounds used, their solubility in water and systemicity. In other words, to reach a successful and optimized blanching, the method must be the least aggressive as possible to sensory and nutritional characteristics but must be as effective as possible to remove antinutritional compounds, chemicals, and microorganisms at maximum. It should be reminded that a gentler processing will be enough for a good quality raw material, what must yield products of superior attributes.

Leaching is the main responsible for removing antinutrients and pesticides, being necessary an immersion blanching to occur. For microbial reduction, steam blanching is also effective, since heat is the primary factor in destructing vegetative cells. Other undesirable substances can (or probably can) be removed by blanching or at least by cooking, as in the case of amylase inhibitors (Chau and Cheung 1997; Gamel et al. 2006), momordicosides (Donya et al. 2007), caffeine (Vuong et al. 2013), lectins (Petres et al. 1990), and mycotoxins (Siwela et al. 2011). However, the information found was not enough to give a broader perspective of the effect of blanching on their removal from foods.

References

Alexandre EMC, Santos-Pedro DM, Brandão TRS, Silva CLM (2011) Study on thermosonication and ultraviolet radiation processes as an alternative to blanching for some fruits and vegetables. Food Bioprocess Technology 4:1012–1019. Doi:10.1007/s11947-011-0540-8

Almazan AM (1995) Antinutritional factors in sweetpotato greens. J Food Compos Anal 8:363–368

Almazan AM, Begum F (1996) Nutrients and antinutrients in peanut greens. J Food Compos Anal 9:375–383. Doi:10.1006/jfca.1996.0043

Amarowicz R (2007) Tannins: the new natural antioxidants? Eur J Lipid Sci Technol 109:549–551. Doi:10.1002/ejlt.200700145

Ani JC, Inyang UE, Udoidem I (2015) Effect of concentration of debittering agent on the mineral, vitamin and phytochemical contents of *Lasianthera africana* leafy vegetable. Afr J Food Sci 9:194–199. Doi:10.5897/AJFS2014.1209

Arcoverde JHV, Carvalho ADS, de Almeida Neves FP et al (2014) Screening of Caatinga plants as sources of lectins and trypsin inhibitors. Nat Prod Res 28:1297–1301. Doi:10.1080/14786419. 2014.900497

Bajwa U, Sandhu KS (2014) Effect of handling and processing on pesticide residues in food- A review. J Food Sci Technol 51:201–220. Doi:10.1007/s13197-011-0499-5

Bhat S, Sharma HK (2016) Combined effect of blanching and sonication on quality parameters of bottle gourd (*Lagenaria siceraria*) juice. Ultrason Sonochem 33:182–189. Doi:10.1016/j. ultsonch.2016.04.014

Bonnechère A, Hanot V, Jolie R et al (2012) Effect of household and industrial processing on levels of five pesticide residues and two degradation products in spinach. Food Control 25:397–406. Doi:10.1016/j.foodcont.2011.11.010

Bouvard V, Loomis D, Guyton KZ et al (2015) Carcinogenicity of consumption of red and processed meat. Lancet Oncol 16:1599–1600. Doi:10.1016/S1470-2045(15)00444-1

Bradbury JH, Hammer B, Nguyen T et al (1985) Protein quantity and quality and trypsin inhibitor content of sweet potato cultivars from the highlands of Papua New Guinea. J Agric Food Chem 33:281–285

Chamberlain K, Evans AA, Bromilow RH (1996) 1-Octanol/water partition coefficient (Kow) and pKa for ionisable pesticides measured by apH-Metric method. Pestic Sci 47:265–271. Doi:10. 1002/(SICI)1096-9063(199607)47:3<265:AID-PS416>3.0.CO;2-F

Chan TYK (2011) Vegetable-borne nitrate and nitrite and the risk of methaemoglobinaemia. Toxicol Lett 200:107–108. Doi:10.1016/j.toxlet.2010.11.002

Chau C-F, Cheung PC-K (1997) Effect of various processing methods on antinutrients and in vitro digestibility of protein and starch of two Chinese indigenous legume seeds. J Agric Food Chem 45:4773–4776. Doi:10.1021/jf970504p

Chung SWC, Tran JCH, Tong KSK et al (2011) Nitrate and nitrite levels in commonly consumed vegetables in Hong Kong. Food Addit Contam Part B 4:34–41. Doi:10.1080/19393210.2011. 557784

Crossey LJ (1991) Thermal degradation of aqueous oxalate species. Geochim Cosmochim Acta 55:1515–1527. Doi:10.1016/0016-7037(91)90124-N

Dao L, Friedman M (1994) Chlorophyll, chlorogenic acid, glycoalkaloid, and protease inhibitor content of fresh and green potatoes. J Agric Food Chem 42:633–639. Doi:10.1021/ jf00039a006

Dimenstein L, Lisker N, Kedar N, Levy D (1997) Changes in the content of steroidal glycoalkaloids in potato tubers grown in the field and in the greenhouse under different conditions of light, temperature and daylength. Physiol Mol Plant Pathol 50:391–402. Doi:10. 1006/pmpp.1997.0098

DiPersio PA, Kendall PA, Yoon Y, Sofos JN (2007) Influence of modified blanching treatments on inactivation of *Salmonella* during drying and storage of carrot slices. Food Microbiol 24:500–507. Doi:10.1016/j.fm.2006.09.004

Donya A, Hettiarachchy N, Liyanage R et al (2007) Effects of processing methods on the proximate composition and momordicosides K and L content of bitter melon vegetable. J Agric Food Chem 55:5827–5833. Doi:10.1021/jf070428i

Dubrow R, Darefsky AS, Park Y et al (2010) Dietary components related to *N*-nitroso compound formation: a prospective study of adult glioma. Cancer Epidemiol Biomark Prev 19:1709–1722. Doi:10.1158/1055-9965.EPI-10-0225

Duval A, Avérous L (2016) Characterization and physicochemical properties of condensed tannins from acacia catechu. J Agric Food Chem 64:1751–1760. Doi:10.1021/acs.jafc.5b05671

Elzbieta R (2012) The effect of industrial potato processing on the concentrations of glycoalkaloids and nitrates in potato granules. Food Control 28:380–384. Doi:10.1016/j.foodcont.2012.04.049

Farrant JM, Traill Z, Conlon C et al (1996) Pigbel-like syndrome in a vegetarian in Oxford. Gut 39:336–337. Doi:10.1136/gut.39.2.336

Finizio A, Vighi M, Sandroni D (1997) Determination of n-octanol/water partition coefficient (Kow) of pesticide critical review and comparison of methods. Chemosphere 34:131–161. Doi:10.1016/S0045-6535(96)00355-4

Foster JW (1993) The acid tolerance response of *Salmonella typhimurium* involves transient synthesis of key acid shock proteins. J Bacteriol 175:1981–1987

Friedman M (2006) Potato glycoalkaloids and metabolites: Roles in the plant and in the diet. J Agric Food Chem 54:8655–8681. Doi:10.1021/jf061471t

Friedman M (2015) Chemistry and anticarcinogenic mechanisms of glycoalkaloids produced by eggplants, potatoes, and tomatoes. J Agric Food Chem 63:3323–3337. Doi:10.1021/acs.jafc.5b00818

Friedman M, Roitman JN, Kozukue N (2003) Glycoalkaloid and calystegine contents of eight potato cultivars. J Agric Food Chem 51:2964–2973. Doi:10.1021/jf021146f

Gamel TH, Linssen JP, Mesallam AS et al (2006) Seed treatments affect functional and antinutritional properties of amaranth flours. J Sci Food Agric 86:1095–1102. Doi:10.1002/jsfa.2463

Gangolli SD, van den Brandt PA, Feron VJ et al (1994) Nitrate, nitrite and N-nitroso compounds. Eur J Pharmacol Environ Toxicol Pharmacol 292:1–38. Doi:10.1016/0926-6917(94)90022-1

Gaugler M, Grigsby WJ (2009) Thermal degradation of condensed tannins from Radiata pine bark. J Wood Chem Technol 29:305–321. Doi:10.1080/02773810903165671

Gómez MJ, Herrera S, Solé D et al (2011) Automatic searching and evaluation of priority and emerging contaminants in wastewater and river water by stir bar sorptive extraction followed by comprehensive two-dimensional gas chromatography-time-of-flight mass spectrometry. Anal Chem 83:2638–2647. Doi:10.1021/ac102909g

Gómez PL, Alzamora SM, Castro MA, Salvatori DM (2010) Effect of ultraviolet-C light dose on quality of cut-apple: Microorganism, color and compression behavior. J Food Eng 98:60–70. Doi:10.1016/j.jfoodeng.2009.12.008

Hildebrand DF, Hamilton-Kemp TR, Loughrin JH et al (1990) Lipoxygenase 3 reduces hexanal production from soybean seed homogenates. J Agric Food Chem 38:1934–1936

Hill RH, Head SL, Baker S et al (1995) Pesticide residues in urine of adults living in the United States: reference range concentrations. Environ Res 71:99–108. Doi:10.1006/enrs.1995.1071

Huan Z, Xu Z, Jiang W et al (2015) Effect of Chinese traditional cooking on eight pesticides residue during cowpea processing. Food Chem 170:118–122. Doi:10.1016/j.foodchem.2014.08.052

Jabbar S, Abid M, Hu B et al (2014) Quality of carrot juice as influenced by blanching and sonication treatments. LWT-Food Sci Technol 55:16–21. Doi:10.1016/j.lwt.2013.09.007

Jakszyn P, Gonzalez C-A (2006) Nitrosamine and related food intake and gastric and oesophageal cancer risk: a systematic review of the epidemiological evidence. World J Gastroenterol 12:4296–4303

Jaworska G (2005) Nitrates, nitrites, and oxalates in products of spinach and New Zealand spinach. Food Chem 93:395–401. Doi:10.1016/j.foodchem.2004.09.035

Kaneko H, Miyamoto J (2001) Pyrethroid chemistry and metabolism. In: Krieger RI, Krieger WC (eds) Handbook of pesticide toxicology, 2nd edn. Academic Press, San Diego, pp 1263–1288

Keikotlhaile BM, Spanoghe P, Steurbaut W (2010) Effects of food processing on pesticide residues in fruits and vegetables: a meta-analysis approach. Food Chem Toxicol 48:1–6. Doi:10.1016/j.fct.2009.10.031

Khan MS, Munir I, Khan I (2013) The potential of unintended effects in potato glycoalkaloids. Afr J Biotechnol 12:754–766. Doi:10.5897/AJBX11.025

Kim NH, Lee NY, Kim SH et al (2015) Optimization of low-temperature blanching combined with calcium treatment to inactivate *Escherichia coli* O157:H7 on fresh-cut spinach. J Appl Microbiol 119:139–148. Doi:10.1111/jam.12815

Kim S-W, Abd El-Aty AM, Rahman MM, et al (2014) Detection of pyridaben residue levels in hot pepper fruit and leaves by liquid chromatography-tandem mass spectrometry: effect of household processes. Biomed Chromatogr n/a–n/a. Doi:10.1002/bmc.3383

Korus A, Lisiewska Z, Słupski J, Gębczyński P (2011) Retention of oxalates in frozen products of three brassica species depending on the methods of freezing and preparation for consumption. Int J Refrig 34:1527–1534. Doi:10.1016/j.ijrefrig.2011.05.009

Kumar V, Sinha AK, Makkar HPS, Becker K (2010) Dietary roles of phytate and phytase in human nutrition: a review. Food Chem 120:945–959. Doi:10.1016/j.foodchem.2009.11.052

Kwon H, Kim T-K, Hong S-M et al (2015) Effect of household processing on pesticide residues in field-sprayed tomatoes. Food Sci Biotechnol 24:1–6. Doi:10.1007/s10068-015-0001-7

Lachman J, Hamouz K, Musilová J et al (2013) Effect of peeling and three cooking methods on the content of selected phytochemicals in potato tubers with various colour of flesh. Food Chem 138:1189–1197. Doi:10.1016/j.foodchem.2012.11.114

Leszczyńska T, Filipiak-Florkiewicz A, Cieślik E et al (2009) Effects of some processing methods on nitrate and nitrite changes in cruciferous vegetables. J Food Compos Anal 22:315–321. Doi:10.1016/j.jfca.2008.10.025

Liener IE, Kakade ML (1969) Protease inhibitors. In: Liener IE (ed) Toxic constituents of plant foodstuffs. New York, pp 8–68

Lisiewska Z, Korus A, Kmiecik W, Gebczyński P (2006) Effect of maturity stage on the content of ash components in raw and preserved grass pea (*Lathyrus sativus* L.) seeds. Int J Food Sci Nutr 57:39–45. Doi:10.1080/09637480500515420

Lv YC, Song HL, Li X et al (2011) Influence of blanching and grinding process with hot water on beany and non-beany flavor in soymilk. J Food Sci 76:20–25. Doi:10.1111/j.1750-3841.2010.01947.x

Mäder J, Rawel H, Kroh LW (2009) Composition of Phenolic Compounds and Glycoalkaloids α-Solanine and α-Chaconine during Commercial Potato Processing. J Agric Food Chem 57:6292–6297. Doi:10.1021/jf901066k

Marshall DL, Dickson JS, Nguyen NH (2016) Ensuring food safety in insect based foods: mitigating microbiological and other foodborne hazards. In: Dossey AT, Morales-Ramos JA, Rojas MG (eds) Insects as sustainable food ingredients. Academic Press, San Diego, pp 223–253

Montagnac JA, Davis CR, Tanumihardjo SA (2009) Processing techniques to reduce toxicity and antinutrients of cassava for use as a staple food. Compr Rev Food Sci Food Saf 8:17–27. Doi:10.1111/j.1541-4337.2008.00064.x

Mosha TC, Gaga HE (1999) Nutritive value and effect of blanching on the trypsin and chymotrypsin inhibitor activities of selected leafy vegetables. Plant Foods Hum Nutr 54:271–283. Doi:10.1023/A:1008157508445

Mozzoni LA, Chen P, Morawicki RO et al (2009) Quality attributes of vegetable soybean as a function of boiling time and condition. Int J Food Sci Technol 44:2089–2099. Doi:10.1111/j.1365-2621.2009.02038.x

Muraleedharan K, Kripa S (2014) DSC kinetics of the thermal decomposition of copper(II) oxalate by isoconversional and maximum rate (peak) methods. J Therm Anal Calorim 115:1969–1978. Doi:10.1007/s10973-013-3366-y

Murugkar DA (2015) Effect of different process parameters on the quality of soymilk and tofu from sprouted soybean. J Food Sci Technol 52:2886–2893. Doi:10.1007/s13197-014-1320-z

Nambisan B, Sundaresan S (1985) Effect of processing on the cyanoglucoside content of cassava. J Sci Food Agric 36:1197–1203. Doi:10.1002/jsfa.2740361126

Nkafamiya II, Oseameahon SA, Modibbo UU, Haggai D (2010) Vitamins and effect of blanching on nutritional and anti-nutritional values of non-conventional leafy vegetables. Afr J Food Sci 4:335–341

Noonan SC, Savage GP (1999) Oxalate content of foods and its effect on humans. Asia Pac J Clin Nutr 8:64–74. Doi:10.1046/j.1440-6047.1999.00038.x

Nwosu JN (2010) Effect of soaking, blanching and cooking on the anti-nutritional properties of asparagus bean (*Vigna sesquipedis*) flour. Nat Sci 8:163–167

Oboh G (2005) Effect of some post-harvest treatments on the nutritional properties of *Cnidoscolus acontifolus* leaf. Pakistan J Nutr 4:226–230. Doi:10.3923/pjn.2005.226.230

Oboh G, Ekperigin MM, Kazeem MI (2005) Nutritional and haemolytic properties of eggplants (*Solanum macrocarpon*) leaves. J Food Compos Anal 18:153–160. Doi:10.1016/j.jfca.2003.12.013

Oner ME, Walker PN (2011) Shelf-life of near-aseptically packaged refrigerated potato strips. LWT-Food Sci Technol 44:1616–1620. Doi:10.1016/j.lwt.2011.02.003

Pao S, Ettinger MR, Khalid MF et al (2008) Microbiological quality of frozen "edamame" (vegetable soybean). J Food Saf 28:300–313. Doi:10.1111/j.1745-4565.2008.00121.x

Patel B, Schutte R, Sporns P et al (2002) Potato glycoalkaloids adversely affect intestinal permeability and aggravate inflammatory bowel disease. Inflamm Bowel Dis 8:340–346. Doi:10.1097/00054725-200209000-00005

Pęksa A, Gołubowska G, Aniołowski K et al (2006) Changes of glycoalkaloids and nitrate contents in potatoes during chip processing. Food Chem 97:151–156. Doi:10.1016/j.foodchem. 2005.03.035

Petres J, Senkalszky-Akos E, Czukor B (1990) Inactivation of trypsin inhibitor, lectin and urease in soybean by hydrothermal treatment. Nahrung 34:905–913

Phillips BJ, Hughes JA, Phillips JC et al (1996) A study of the toxic hazard that might be associated with the consumption of green potato tops. Food Chem Toxicol 34:439–448. Doi:10.1016/0278-6915(96)87354-6

Radwan MA, Abu-Elamayem MM, Shiboob MH, Abdel-Aal A (2005) Residual behaviour of profenofos on some field-grown vegetables and its removal using various washing solutions and household processing. Food Chem Toxicol 43:553–557. Doi:10.1016/j.fct.2004.12.009

Roddick JG, Weissenberg M, Leonard AL (2001) Membrane disruption and enzyme inhibition by naturally-occurring and modified chacotriose-containing Solanum steroidal glycoalkaloids. Phytochemistry 56:603–610. Doi:10.1016/S0031-9422(00)00420-9

Rytel E (2013) Effect of technological factors on glycoalkaloids and nitrates content in dehydrated potato. Ital J Food Sci 25:142–148

Rytel E (2012a) Changes in glycoalkaloid and nitrate content in potatoes during dehydrated dice processing. Food Control 25:349–354. Doi:10.1016/j.foodcont.2011.10.053

Rytel E (2012b) Changes in the levels of glycoalkaloids and nitrates after the dehydration of cooked potatoes. Am J Potato Res 89:501–507. Doi:10.1007/s12230-012-9273-0

Rytel E, Gołubowska G, Lisińska G et al (2005) Changes in glycoalkaloid and nitrate contents in potatoes during French fries processing. J Sci Food Agric 85:879–882. Doi:10.1002/jsfa.2048

Rytel E, Tajner-Czopek A, Aniołowska M, Hamouz K (2013) The influence of dehydrated potatoes processing on the glycoalkaloids content in coloured-fleshed potato. Food Chem 141:2495–2500. Doi:10.1007/s00217-014-2163-6

Sallau AB, Mada SB, Ibrahim S, Ibrahim U (2012) Effect of boiling, simmering and blanching on the antinutritional content of Moringa oleifera leaves. Int J Food Nutr Saf 2:1–6

Savage GP, Vanhanen L, Mason SM, Ross AB (2000) Effect of cooking on the soluble and insoluble oxalate content of some New Zealand foods. J Food Compos Anal 13:201–206. Doi:10.1006/jfca.2000.0879

Shi J, Arunasalam K, Yeung D et al (2004) Saponins from edible legumes: chemistry, processing, and health benefits. J Med Food 7:67–78. Doi:10.1089/109662004322984734

Shi J, Xue SJ, Ma Y et al (2009) Kinetic study of saponins B stability in navy beans under different processing conditions. J Food Eng 93:59–65. Doi:10.1016/j.jfoodeng.2008.12.035

Siener R, Hönow R, Seidler A et al (2006) Oxalate contents of species of the polygonaceae, amaranthaceae and chenopodiaceae families. Food Chem 98:220–224. Doi:10.1016/j. foodchem.2005.05.059

Sinden SL, Deahl KL, Aulenbach BB (1976) Effect of glycoalkaloids and phenolics on potato flavor. J Food Sci 41:520–523. Doi:10.1111/j.1365-2621.1976.tb00661.x

Siwela AH, Mukaroa KJ, Nziramasanga N (2011) Aflatoxin Carryover during large scale peanut butter production. Food Nutr Sci 2:105–108. Doi:10.4236/fns.2011.22014

Smith DB, Roddick JG, Jones JL (1996) Potato glycoalkaloids: Some unanswered questions. Trends Food Sci Technol 7:126–131. Doi:10.1016/0924-2244(96)10013-3

Somsub W, Kongkachuichai R, Sungpuag P, Charoensiri R (2008) Effects of three conventional cooking methods on vitamin C, tannin, myo-inositol phosphates contents in selected Thai vegetables. J Food Compos Anal 21:187–197. Doi:10.1016/j.jfca.2007.08.002

Sucha L, Tomsik P (2016) The steroidal glycoalkaloids from solanaceae: toxic effect, antitumour activity and mechanism of action. Planta Med 82:379–387. Doi:10.1055/s-0042-100810

Tadeo JL, Sánchez-Brunete C, González L (2008) Pesticides: classification and properties. In: Tadeo JL (ed) Analysis of pesticides in food and environmental samples. CRC Press, Boca Raton, pp 1–34

Tajner-Czopek A, Rytel E, Kita A et al (2012) The influence of thermal process of coloured potatoes on the content of glycoalkaloids in the potato products. Food Chem 133:1117–1122. Doi:10.1016/j.foodchem.2011.10.015

Tajner-Czopek A, Rytel E, Aniołowska M, Hamouz K (2014) The influence of French fries processing on the glycoalkaloid content in coloured-fleshed potatoes. Eur Food Res Technol 238:895–904. Doi:10.1007/s00217-014-2163-6

Tricker AR, Preussmann R (1991) Carcinogenic N-nitrosamines in the diet: occurrence, formation, mechanisms and carcinogenic potential. Mutat Res Toxicol 259:277–289. Doi:10.1016/0165-1218(91)90123-4

Tunçel G, Nout MJR, Brimer L (1998) Degradation of cyanogenic glycosides of bitter apricot seeds (*Prunus armeniaca*) by endogenous and added enzymes as affected by heat treatments and particle size. Food Chem 63:65–69. Doi:10.1016/S0308-8146(97)00217-3

Uusiku NP, Oelofse A, Duodu KG et al (2010) Nutritional value of leafy vegetables of sub-Saharan Africa and their potential contribution to human health: a review. J Food Compos Anal 23:499–509. Doi:10.1016/j.jfca.2010.05.002

van Ravenzwaay B, Leibold E (2004) A comparison between in vitro rat and human and in vivo rat skin absorption studies. Hum Exp Toxicol 23:421–430. Doi:10.1191/0960327104ht471oa

Vanderjagt DJ, Freiberger C, Vu HTN et al (2000) The trypsin inhibitor content of 61 wild edible plant foods of Niger. Plant Foods Hum Nutr 55:335–346. Doi:10.1023/A:1008136100545

Vandeweyer D, Lenaerts S, Callens A, Van Campenhout L (2017) Effect of blanching followed by refrigerated storage or industrial microwave drying on the microbial load of yellow mealworm larvae (*Tenebrio molitor*). Food Control 71:311–314. Doi:10.1016/j.foodcont.2016.07.011

Vetter J (2000) Plant cyanogenic glycosides. Toxicon 38:11–36. Doi:10.1016/S0041-0101(99)00128-2

Vighi M, Di Guardo A (1995) Preditive approaches for the evaluation of pesticide exposure. In: Vighi M, Funari E (eds) Pesticide risk in groundwater. Lewis Publishers Inc., Boca Raton, pp 73–100

Villalobos MC, Serradilla MJ, Martín A et al (2016) Evaluation of different drying systems as an alternative to sun drying for figs (*Ficus carica* L). Innov Food Sci Emerg Technol 36:156–165. Doi:10.1016/j.ifset.2016.06.006

Vuong QV, Golding JB, Nguyen MH, Roach PD (2013) Preparation of decaffeinated and high caffeine powders from green tea. Powder Technol 233:169–175. Doi:10.1016/j.powtec.2012.09.002

Wang N, Lewis MJ, Brennan JG, Westby A (1997) Effect of processing methods on nutrients and anti-nutritional factors in cowpea. Food Chem 58:59–68. Doi:10.1016/S0308-8146(96)00212-9

Weitzberg E, Lundberg JO (2013) Novel aspects of dietary nitrate and human health. Annu Rev Nutr 33:129–159. Doi:10.1146/annurev-nutr-071812-161159

Yadav SK, Sehgal S (2003) Effect of domestic processing and cooking on selected antinutrient contents of some green leafy vegetables. Plant Foods Hum Nutr 58:1–11. Doi:10.1023/B:QUAL.0000040359.40043.4f

Yuan S, Chang SKC, Liu Z, Xu B (2008) Elimination of trypsin inhibitor activity and beany flavor in soy milk by consecutive blanching and ultrahigh-temperature (UHT) processing. J Agric Food Chem 56:7957–7963. Doi:10.1021/jf801039h

Chapter 5
Blanching as an Acrylamide Mitigation Technique

João Luiz Andreotti Dagostin

Abstract Food processing is a fundamental activity applied to extend the shelf-life, to improve stabilization, increase the nutritional value, provide convenience and give special characteristics to foods and foodstuffs. Within the set of thermal technologies applied in food processing, cooking is a step intended to give unique sensory properties, as texture, flavor, color, and taste. During the processes of frying, baking, or roasting for some types of foods—especially carbohydrate-rich foods—Maillard reactions may take place, yielding a pleasant flavor and color compounds. Some pathways of the Maillard reaction chain, however, may produce undesirable substances like acrylamide, furans, and furfurals. Acrylamide in particular concerns researchers and consumers since it is classified as a probable human carcinogen. In this chapter, blanching will be presented as an acrylamide mitigation technology, highlighting some of the most relevant studies involving the method, the effect of time and temperature, and the combination of different technologies to blanching.

Keywords Blanching · Acrylamide · Maillard reaction

Knowing Acrilamide

Acrylamide (prop-2-enamide, C_3H_5NO, CAS# 79-06-01) is a white, colorless, odorless, crystalline water-soluble amide. It is commonly used in the chemical industry as an intermediate in the synthesis of polyacrylamide hydrogels. These hydrogels may find a great range of applications, such as in water treatment as flocculants, in paper and textile industries as a binder, as gel electrophoresis media, for toys production, as soil conditioners, in dye preparation, among others. While

J.L.A. Dagostin (✉)
Graduate Program in Food Engineering, Chemical Engineering Department, Federal University of Paraná, 19011 Francisco H. dos Santos (S/No), Curitiba 81531-980, Paraná, Brazil
e-mail: joaodagostin@hotmail.com

© Springer International Publishing AG 2017   95
F. Richter Reis (ed.), *New Perspectives on Food Blanching*,
DOI 10.1007/978-3-319-48665-9_5

the polymer polyacrylamide is non-toxic, its monomer acrylamide may represent a health concern.

Acrylamide consumption concerns food researchers since the substance is known for its neurotoxicity to humans. The National Toxicology Program (NTP) of the U.S. Department of Health and Human Services (HHS) and the International Agency for Research on Cancer (IARC)—a specialized cancer agency of the World Health Organization (WHO)—classify acrylamide as a probable carcinogen (Group 2A) based on studies involving the administration of the substance in rodents, the consumption of acrylamide-containing food by humans, or the human exposure to acrylamide. After taking doses of acrylamide in drinking water, there was an incidence of thyroid, mammary, and pituitary cancer (among other tumor types) in rats (Johnson et al. 1986; Friedman et al. 1995). The real impact of the acrylamide intake on humans is still uncertain. However, epidemiological studies link its consumption with breast and oral cavity cancers, renal cells cancer, endometrial cancer and ovarian cancers, as well documented by Hogervorst et al. (2010). To better know about the analysis of acrylamide, its formation and health effects, it is recommended the work of Gökmen (2016).

The presence of acrylamide in raw, *in natura* or low processed foods is not likely to occur. It was in the early 2000s when the first reports about acrylamide presence and formation in foodstuffs were published by Swedish scientists (Tareke et al. 2000, 2002). After this pioneer study, the researchers could verify that different cooking methods, the cooking time, and the cooking temperature of food sources were parameters that may interfere on the acrylamide production in a higher or lower extent. After baking pre-fried potatoes for assessing the temperature influence, they found that the temperature necessary to significantly increase acrylamide content should be >100 °C (for a 19 min assay). However, it seems that a significant change in the acrylamide formation is achieved when 120 °C or above is applied in the thermal process. For more details about the course of discovering acrylamide in processed foods at the earlier stages, you may take a look at the work of Törnqvist (2005).

After the discovery of acrylamide in processed foods, a large number of studies emerged in order to investigate in what circumstances it would occur and in what extent. The work of Tareke et al. (2002) reveals a first screening of the parameters involved in acrylamide formation. Considering the parameters studied, the authors have detected moderate amounts of acrylamide (5–50 µg/kg) in protein-rich foods (e.g., beef, chicken and fish), while a high content (150–4000 µg/kg) was observed for carbohydrate-rich foods (such as potatoes, beetroot, commercial crispbread, and potato products). From the different cooking methods used—heating in a frypan, boiling, and microwave heating—boiling was the only one that presented insignificant levels of acrylamide formation (<5 µg/kg). Since then, there were advances in finding which foodstuffs tend to present higher or lower amounts of acrylamide as a consequence of their composition, processing time, and degree of thermal processing.

Carbohydrate-rich foods are the main sources of acrylamide formation. The complete mechanism of acrylamide formation in foods has not been fully elucidated yet. Most of the mechanisms proposed involve the degradation of free asparagine when reducing sugars are available during a Maillard reaction in thermal treatments as frying, baking, or roasting. The Maillard reaction actually involves a sequence of reactions starting majorly from reducing carbohydrates and free amino groups of amino acids and/or proteins. This reaction may develop desirable characteristics to the food, as flavor and golden, brown, or caramel color. However within the reaction chain some other chemical routes may be taken and undesirable products can be formed, such as acrylamide, furans, and (hydroxymethyl)furfural (Friedman 2015). The pathways to acrylamide formation from the postulated chemical mechanisms involving a reducing sugar and asparagine are shown in Fig. 5.1. The asparagine-reducing sugar pathways are most likely to be the main mechanisms of acrylamide formation in foods. However, other mechanisms, involving thermolytic pathways—via the Strecker aldehyde, acrolein, and acrylic acid—and the decarboxylation of asparagine—with the aid of a specific enzyme and a cofactor in an aqueous system—were also proposed by Yaylayan and Stadler (2005) and Granvogl et al. (2004), respectively. These should be responsible for acrylamide formation in specific cases and in a lesser extent than the asparagine-reducing sugar routes.

Although cysteine, methionine, serine, and threonine are also precursors of acrylamide when in the presence of reducing sugars, the yields of acrylamide from these reactions are much lower than those resulting from asparagine (Belitz et al. 2009). In fact, not even the presence of reducing sugars is necessary to transform asparagine into acrylamide. Different molecules carrying carbonyl groups are

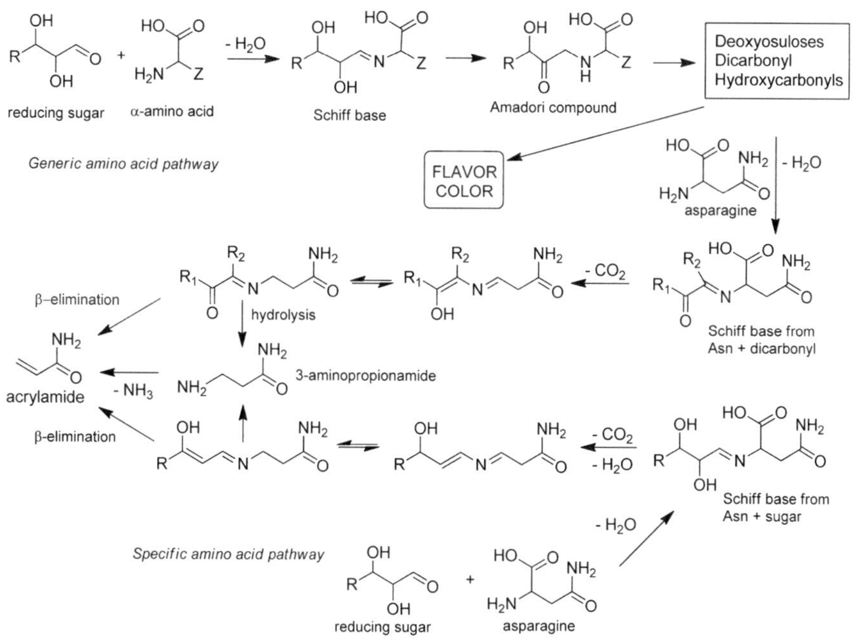

Fig. 5.1 Postulated mechanisms of acrylamide formation from asparagine-reducing sugar interactions (Parker et al. 2012)

capable to react with asparagine at a higher or lower rate to obtain acrylamide. Some of them include vitamins, lipid oxidation products, aldehydes, hydroxymethylfurfural, and amino acids (Zamora and Hidalgo 2008; Gökmen et al. 2012; Hamzalıoğlu and Gökmen 2012). However, reducing sugars are largely available in food systems and represent the major carbonyl precursor of acrylamide formation (Capuano 2016).

Acrylamide Mitigation Techniques

Thermal processes involving low temperatures or water (as boiling or steam cooking) to transfer energy to foods are known to reduce Maillard reactions and consequently decrease acrylamide formation. Despite the proved existence of acrylamide in certain food types, nowadays there is enough information relating some techniques and precautions in food processing which may disfavor acrylamide production. As you may expect, blanching is one of the successful techniques used to reduce acrylamide content in foods. Before introducing this subject, a brief review about the mitigation techniques used for this purpose will be presented since it is a common strategy to perform the application of combined methods as means of increase acrylamide mitigation efficiency and reduce technological and sensorial problems. Based on the classifications found in the works of Xu et al. (2016), Morales et al. (2008), Anese et al. (2010) and others, some of the main acrylamide mitigation techniques were here classified as follows:

1. **Use of raw material low in acrylamide precursors, mainly reducing sugars and asparagine**. This could be achieved by partially substituting the raw material, choosing cultivars low in reducing sugars and/or free asparagine. The addition of glucose and specially fructose to a baked potato model system, for example, was found to increased acrylamide formation in 160% and 460%, respectively (Rydberg et al. 2003).
2. **Controlling parameters of the process or food characteristics**. Some of the main parameters to be changed in the blanching process are temperature and time. Food pH could also be changed to control acrylamide generation (Rydberg et al. 2003; Vinci et al. 2012). Moisture and water activity also play important roles in Maillard and acrylamide formation. Higher water availability and water content generally decrease acrylamide production. The formation of acrylamide in food usually start at a water activity of 0.8, with a peak of production at 0.4. Also, food crusts having moisture contents ranging from 2% to 6% may present higher final acrylamide amounts (Mesias and Morales 2016). Both water activity and moisture content are variables linked to the food type and processing conditions when evaluating the final acrylamide level. Blanching is a simple technology that can be applied in acrylamide reduction. Immersive and steam blanching are restricted to certain food types: low processed solid foods. Using blanching in vegetables prior to cooking helps to reduce acrylamide precursors

while protecting the tissue against enzymatic activity. The use of such technology will be further detailed. Food shape and size are other parameters affecting acrylamide formation. Thinner food samples should exhibit a higher drying rate, leading to a higher acrylamide content (Mesias and Morales 2016). Foods with a larger superficial area are also able to form a larger crust during thermal treatments. For a larger crust, the water loss, the Maillard precursor's formation and the thermal exposure are higher, ultimately leading to a higher acrylamide formation.

3. **Use of specific additives, competitive inhibitors, or enzymes**. There are several works reporting a range of additives meant to mitigate acrylamide formation. As different additive types may exert different inhibitory mechanisms, care should be taken when evaluating the models of acrylamide formation. Some of the substances found to mitigate acrylamide production include phenolic compounds and other antioxidants, as vitamin E, tert-butylhydroquinone (Li et al. 2012), syntetic BHA and BHT, ferrulic acid, vitamin C (Ou et al. 2008; Urbančič et al. 2014) and naringenin (Cheng et al. 2009). Besides, plant antioxidant mixes and plant extracts naturally containing phenolics were also verified as good alternatives to the purified substances. Successful examples are the insertion of grape extract (Xu et al. 2015), rosemary aqueous extract (Hedegaard et al. 2008; Urbančič et al. 2014), green tea extract, and antioxidants of oregano (Kotsiou et al. 2010) and bamboo (Zhang and Zhang 2008; Li et al. 2012) in food systems. Other additives used for the purpose of reducing acrylamide formation include citric acid (Jung et al. 2003), sodium bisulphite (Casado et al. 2010), ascorbic acid (Chen and Gu 2014), sodium erythorbate (Li et al. 2012), tyrosol, oleuropein, *p*-hydroxyphenylacetic acid (Kotsiou et al. 2010), *p*-coumaric acid (Xu and An 2016), among others.

Since the main route of acrylamide formation involves the reactions between reducing sugars and asparagine, the application of asparaginase has been studied since 2003 (Zyzak et al. 2003) as a solution to mitigate acrylamide formation. Despite that the use of asparaginase in food processes indicates positive results, its application in an industrial scale may be limited to the process (batch in most of the cases), food type, pH and temperature, which are factors influencing enzyme reaction rate and its destruction. Nowadays there are microbial asparaginases commercially available for use in food industries, including those from *Aspergillus oryzae* and *Aspergillus niger*. However, studies indicate the production of asparaginase from other sources like an *Escherichia coli* transformed by a *Thermococcus zilligii* gene (Zuo et al. 2015), the marine bacteria *Bacillus tequilensis* PV9W (Shakambari et al. 2016), *Aquabacterium* sp. A7-Y genes (Sun et al. 2016), *Bacillus subtilis* KDPS1 (Sanghvi et al. 2016), *Pseudomonas oryzihabitans* (Bhagat et al. 2016), and so many others. More information about the use of asparaginase to reduce acrylamide formation in foods may be found in a review published by Xu et al. (2016).

4. **Application of post-processes interventions**. This category commonly involves complex or costly techniques to be applied in food systems to remove, consume, or degrade acrylamide—especially when applied exclusively to this function. Rivas-Jimenez et al. (2016) for example, verified that *Lactobacillus reuteri* NRRL 14171 and especially *Lactobacillus casei* Shirota could successfully remove acrylamide under simulated gastrointestinal conditions using a dynamic system. Another study presents the possibility of removing acrylamide of low water activity finished products by applying a hydration step followed by vacuum-drying (Anese et al. 2010). The authors assessed the use of vacuum treatments at different combinations of pressure, temperature, and time in the acrylamide removal of commercial biscuits and potato chips. They verified that the acrylamide removal was achieved only in those samples previously hydrated to water activity values higher than 0.83.

Other strategies include: the use of nucleophiles as amino and sulfhydryl groups to react with acrylamide (Adams et al. 2010; Hidalgo et al. 2011); the application of a supercritical CO_2 extraction in roasted coffee (Banchero et al. 2013); the use of microbial extracts containing acrylamidases in coffee (Cha 2013); and the storage of rye crispbread under controlled temperature and moisture (Mustafa et al. 2008).

Blanching as an Acrylamide Mitigation Technique

Time and Temperature Dependence

Using pure water to control acrylamide formation seems to be a time and temperature-dependent process. In any case, the immersion of food in water at ambient (soaking) and high temperature (blanching) leads to basically the same inhibition mechanism: the leaching of acrylamide precursors (Burch et al. 2008). However, the process involving a higher temperature is a key factor when thinking in efficiency. In a shorter period, a greater amount of reducing sugars and asparagine are removed from the food matrix when blanching is used instead of a simple soaking. It is important to note that there is not a temperature range limiting what is soaking and what is blanching. As blanching is intended to inactivate different enzymes, there is a great range of Time × Temperature sets that can be used to achieve the final goal. So it is not hard to find documents involving the terms "soaking" or "blanching" when the immersion is made in water of low or mild temperature. In this chapter, it will be named as "soaking" those processes occurring up to 50 °C. From this temperature on, it will be considered the existence of a blanching process.

When soaking potato slices prior to frying, Pedreschi et al. (2004) verified a 32% loss of original sucrose content and no asparagine variation. In the case of using blanching, a 76 and 68% decrease (on average, for different temperature/time sets) was found for the respective components. Even taking into account a greater precursors loss as pointed above, the whole process can be affected if there is not enough blanching time. While blanching for 2–8 min (90 °C) did not cause a substantial decrease in acrylamide formation, applying 50 °C/70 min and 70 °C/40 min led to a 97 and 91% reduction, respectively, on the acrylamide content of potato slices after frying. A similar trend was found by Mestdagh et al. (2008b), who studied the Time × Temperature profile of blanching on the acrylamide formation of French fries and potato crisps. According to the equations generated, a reduction near 65% and 95% in acrylamide content would be achieved for the products, respectively, when carrying a blanching at 70 °C/10–30 min.

The use of mild to high temperatures (50–90 °C) in blanching should be enough to attain a proper acrylamide mitigation. Using lower blanching temperatures would, of course, require longer processing times. Regarding only this statement, you may ask yourself: "Why not using a higher blanching temperature at once?". But choosing the best parameters for a conventional blanching prior to thermal processing is not an easy task. In Chap. 2 you have seen that the use of different temperatures and times may affect different quality parameters on the final product. For the reduction of acrylamide content, this is not different. Acrylamide yield should be considered as another quality parameter for the whole process if there is an aim of controlling it. Tables 5.1 and 5.2 show some of the best or most relevant blanching parameters used in studies involving the acrylamide mitigation in food products. The references in both tables represent the majority of the papers found for the field until the middle of 2016.

It is important to mention here that searching the literature for processes where blanching was used in different food types as an acrylamide mitigation technique was a frustrating task. The vast majority of studies describing them involve the use of regular potatoes as food matrix. The only exception found involves the frying of sweet potatoes (Truong et al. 2014). It is still uncertain, thus, the effect of blanching on the posterior acrylamide occurrence in different food matrices.

Analyzing those processes with no combined methods during or after the blanching step, some conclusions can be drawn when thinking in using hot water to mitigate acrylamide formation. On the one hand, although using water at higher temperatures (85–95 °C) should provide better results in enzyme denaturation efficiency, the time of immersion must be reduced (HTST process, High Temperature Short Time) to avoid the tissue softening, which decreases the final textural quality of fries (Agblor and Scanlon 2000). The high temperature blanching also contributes to a higher oil uptake when frying the product (Mariotti et al. 2015). On the other hand, blanching at low temperatures (50–65 °C) may not represent an optimal choice. If one uses a low thermal processing (LTLT process, Low Temperature Long Time), there is a risk of insufficient heat to destroy endogenous enzymes and a very long time of processing should be satisfied. Unlike the blanching of raw tissues to stabilize pigments, the color development when cooking (fried or roasted) blanched potatoes seems to be improved by longer immersion times (Agblor and Scanlon 2000). Higher color quality is commonly attributed to higher lightness values (L* in CIE L*a*b* color space) when assessing

Table 5.1 Relevant data of blanching studies involving the acrylamide mitigation in French fries

Potato cultivar	Blanching parameters	Combined methods (in sequence)	Thermal post-processing (in sequence)	Acrylamide content (μg/kg)	Acrylamide reduction	Reference
Favorita (8 × 8 mm cross-section strips)	80 °C/10 min	n.p.	- Drying (85 °C/10 min); - Frying (175 °C/~ 5 min)	967	39%	Zuo et al. (2015)
Favorita (8 × 8 mm strips)	80 °C/8 or 15 min	Use of L-asparaginase (10 U/mL) in blanching water	- Drying (85 °C/10 min); - Frying (175 °C/~ 5 min)	275 (8 min) 232 (15 min)	83% (8 min) 85% (15 min)	Zuo et al. (2015)
Covington[a] (90 × 7.5 × 7.5 mm)	95 °C/3 min	- Blanching in 0.5% sodium acid pyrophosphate (62 °C/10 min); - Soaking (21 °C/10 min)	- Drying (65 °C/10 min); - Par-frying (165 °C/1 min); - Freezing (−20 °C); - Frying (165 °C/2–5 min)	0.016–0.058	~ 86%	Truong et al. (2014)
Covington[a] (90 × 7.5 × 7.5 mm)	95 °C/3 min	- Blanching in 0.5% disodium pyrophosphate (62 °C/10 min); - Soaking (21 °C/10 min); - Blanching in 0.4% CaCl$_2$ (62 °C/10 min)	- Air drying (65 °C/10 min); - Par-frying (165 °C/1 min); - Freezing (−20 °C); - Frying (165 °C/2–5 min)	0.006–0.035	~ 93%	Truong et al. (2014)
Asterix; Fontana;		n.p.	- Par-frying (180 °C/1.5 min);	141 (Fontana)	48–53%	Vinci et al. (2010)

(continued)

Table 5.1 (continued)

Potato cultivar	Blanching parameters	Combined methods (in sequence)	Thermal post-processing (in sequence)	Acrylamide content (μg/kg)	Acrylamide reduction	Reference
Lady Olympia (30 × 10 × 10 mm)	70 °C/20 min at 5:1 (water:potato, w/w)		– Freezing (n.a.); – Frying (175 °C/2.5 min)	180 (Lady Olympia) 727 (Asterix)		
Markies; Bintje; Russet Burbank; Maris Piper (30 × 10 × 10 mm)	70 °C/20 min at 5:1 (water:potato, w/w)	n.p.	– Par-frying (180 °C/1.5 min); – Freezing (n.a.); – Frying (175 °C/2.5 min)	61 (Markies) 305 (Bintje) 375 (Russet Burbank) 454 (Maris Piper)	57–59%	Vinci et al. (2010)
Agria; Innovator (30 × 10 × 10 mm)	70 °C/20 min at 5:1 (water:potato, w/w)	n.p.	– Par-frying (180 °C/1.5 min); – Freezing (n.a.); – Frying (175 °C/2.5 min)	211 (Innovator) 341 (Agria)	63–66%	Vinci et al. (2010)
Victoria (30 × 10 × 10 mm)	70 °C/20 min at 5:1 (water:potato, w/w)	n.p.	– Par-frying (180 °C/1.5 min); – Freezing (n.a.); – Frying (175 °C/2.5 min)	73	86%	Vinci et al. (2010)
Nicola (3 × 1 × 1 cm)	86 °C/20 min at 10:1 (water:potato, w/w)	n.p.	– Frying (180 °C/5 min)	~ 400	81%	Mestdagh et al. (2008b)
Bintje (50 × 8 × 8 mm)	75 °C/10 min at 2:1 (water:potato, w/w)	n.p.	– Drying (85 °C/10 min); – Par-frying (175 °C/1 min); – Freezing (−30 °C/2 days); – Frying (175 °C/3 min)	1264	39%	Pedreschi et al. (2008)
Bintje			– Drying (85 °C/10 min);	483	77%	

(continued)

Table 5.1 (continued)

Potato cultivar	Blanching parameters	Combined methods (in sequence)	Thermal post-processing (in sequence)	Acrylamide content (µg/kg)	Acrylamide reduction	Reference
(50 × 8 × 8 mm)	75 °C/10 min at 2:1 (water:potato, w/w)	Soaking (40 °C/20 min) in a 10,000 ASNU/L asparaginase solution at 2:1 (water:potato, w/w)	– Par-frying (175 °C/1 min); – Freezing (−30 °C/2 days); – Frying (175 °C/3 min)			Pedreschi et al. (2008)
Beate (9 × 9 mm cross-section strips)	80 °C/5 min	Fermentation with *Lactobacillus plantarum* NC8 (10^9 CFU/mL, 37 °C/45 and 120 min)	– Pre-frying in palm oil (170 °C/3 min); – Cooling in ambient air (5 min); – Frying in palm oil (170 °C/2.4 min)	~504 (45 min fermentation) ~144 (120 min fermentation)	79% (45 min fermentation) 94% (120 min fermentation)	Baardseth et al. (2006)
Bintje (50 × 8 × 8 mm)	70 °C/45 min at ~83:1 (water:potato, w/w)	n.p.	– Frying (170 °C/8.5 min) until moisture of ~40%	~514	~86%	Pedreschi et al. (2006)
Beate (n.a.)	70 °C/10 min	n.p.	– Drying (20 °C/5 min); – Pre-frying in palm oil (180 °C/1 min); – Freezing (n.a.); – Frying in palm oil (180 °C/3 min)	244	n.a.	Brathen et al. (2005)

n.p. Not performed
n.a. Information or detailing not available
[a]Sweet potato cultivar

Table 5.2 Relevant data of blanching studies involving the acrylamide mitigation in potato chips/crisps

Potato cultivar	Blanching parameters	Combined methods (in sequence)	Thermal post-processing (in sequence)	Acrylamide content (μg/kg)	Acrylamide reduction	Reference
Ditta (2.2 × 40 mm slices)	64 °C/17 min at ~67:1 (water:potato, w/w)	n.p.	Frying (170 °C) until moisture of 2%	3.4 (estimated from a desirability function)	54% (estimated)	Mariotti et al. (2015)
Agria (1.5 mm slices)	83 °C/2.5 min at ~33:1 (water:potato, w/w)	Soaking for 1 min before blanching	Frying (170 °C/5 min) until moisture of 1.5%	~1000	~76%	Shojaee-Aliabadi et al. (2013)
Agria (1.5 mm slices)	75 °C/9 min at ~33:1 (water:potato, w/w)	Soaking for 1 min before blanching	Frying (170 °C/5 min) until moisture of 1.5%	~450	~90%	Shojaee-Aliabadi et al. (2013)
Verdi (2 × 40 mm slices)	85 °C/3.5 min at 2:1 (water: potato, w/w)	Rinsing for 1 min before blanching	Frying in palm oil (170 °C/5 min) until moisture of ~2.0%	1697	17%	Pedreschi et al. (2011)
Verdi (2 × 40 mm slices)	85 °C/3.5 min at 2:1 (water: potato, w/w)	– Rinsing for 1 min before blanching; – Soaking (50 °C/20 min)	Frying in palm oil (170 °C/5 min) until moisture of ~2.0%	790	62%	Pedreschi et al. (2011)
Verdi (2 × 40 mm slices)	85 °C/3.5 min at 2:1 (water: potato, w/w)	– Rinsing for 1 min before blanching; – Immersion in a 10,000 ASNU/L-asparaginase solution (50 °C/20 min)	Frying in palm oil (170 °C/5 min) until moisture of ~2.0%	158	92%	Pedreschi et al. (2011)

(continued)

Table 5.2 (continued)

Potato cultivar	Blanching parameters	Combined methods (in sequence)	Thermal post-processing (in sequence)	Acrylamide content (μg/kg)	Acrylamide reduction	Reference
Verdi (2 × 40 mm slices, t × d)	90 °C/5 min	n.p.	Frying in palm oil (170 °C/5 min) until moisture of ~2.0%	~2600	n.s.r.	Pedreschi et al. (2010)
Verdi (2 × 40 mm slices, t × d)	90 °C/5 min	Soaking in a 1% NaCl solution (25 °C/5 min)	Frying in palm oil (170 °C/5 min) until moisture of ~2.0%	~960	~62%	Pedreschi et al. (2010)
Saturna; Hulda; SW 91 102 (1.5 mm slices)	80 °C/3 min at 100:1 (water: potato, w/w)	Potatoes were stored for 12 weeks (8 °C)	Frying in rapeseed oil (180 °C/3 min) until moisture of ~2.0%	~1100 (on average for the three potato types)	68–73%	Viklund et al. (2010)
Bintje (1.4 mm slices)	80 °C/1 min at 5:1 (water: potato, w/w)	– Cooling (n.a.); – Soaking (40 °C/15 min) at 5:1 (water:potato, w/w)	Frying (180 °C/2.5 min)	1750	Unblanched samples were not tested	Hendriksen et al. (2009)
Bintje (1.4 mm slices)	80 °C/1 min at 5:1 (water: potato, w/w)	– Cooling (n.a.); – Soaking (40 °C/15 min) in a10500 ASNU/L asparaginase solution at 5:1 (water:potato, w/w)	Frying (180 °C/2.5 min)	710	59%[a] (Unblanched samples were not tested)	Hendriksen et al. (2009)
Bintje (1.5 mm slices)	65 °C/5 min at a 10:1 (water: potato, w/w)	Blanching water containing NaCl (0.05 or 0.1 mol.L^{-1})	Frying (170 °C/3 min)	n.a.	28%[a] (0.05 mol.L^{-1}) 43%[a] (0.1 mol.L^{-1})	Mestdagh et al. (2008a)

(continued)

Table 5.2 (continued)

Potato cultivar	Blanching parameters	Combined methods (in sequence)	Thermal post-processing (in sequence)	Acrylamide content (µg/kg)	Acrylamide reduction	Reference
Bintje (1.5 mm slices)	65 °C/5 min at a 10:1 (water: potato, w/w)	Blanching water containing: – Citric acid or Acetic acid or L-lactic acid (0.025 mol/L); – Glycine or L-lysine (0.05 mol/L)	Frying (170 °C/3 min)	n.a.	98%[a] (Citric acid) 89%[a] (L-lactic acid) 80%[a] (Acetic acid) 63%[a] (Glycine and L-lysine)	Mestdagh et al. (2008a)
Bintje (1.5 mm slices)	65 °C/5 min at a 10:1 (water: potato, w/w)	Blanching water containing sodium acid pyrophosphate (SAPP, 0.05 mol/L)	Frying (170 °C/3 min)	n.a.	98%[a]	Mestdagh et al. (2008a)
Marabel (1.2 mm slices)	70 °C/30 min at 20:1 (water: potato, w/w)	n.p.	Frying (170 °C/1.5 min)	~ 160	98%	Mestdagh et al. (2008b)
Yueyin 2 (25 × 25 × 2.5 mm slices)	85 °C/2.5 min	Blanching water containing: – Cysteine or CaCl$_2$ (0.3%); – NaHSO$_3$ (0.5%)	Frying (150 °C/10 min until moisture of ~ 2.5%	487 (Cysteine) 601 (CaCl$_2$) 2856 (NaHSO$_3$)	91%[a] (Cysteine) 89%[a] (CaCl$_2$) 47%[a] (NaHSO$_3$)	Ou et al. (2008)
Asterix (1.8 mm slices)	80 °C/2 min at 2:1 (water: potato, w/w)	n.p.	Frying in palm oil (170 °C/4 min)	4127	33%	Brathen et al. (2005)
Asterix (1.8 mm slices)	80 °C/2 min at 2:1 (water: potato, w/w)	Use of glycine (0.01 M) in blanching water	Frying in palm oil (170 °C/4 min)	2482	60%	Brathen et al. (2005)

(continued)

Table 5.2 (continued)

Potato cultivar	Blanching parameters	Combined methods (in sequence)	Thermal post-processing (in sequence)	Acrylamide content (μg/kg)	Acrylamide reduction	Reference
Panda (2.2 × 37 mm slices, t × d)	85 °C/3.5 min at 200:1 (water: potato, w/w)	Rinsing for 1 min before blanching	Frying (170 °C/4 min) until moisture of ~ 1.8%	~ 150	75%	Pedreschi et al. (2005)
Asterix (1 mm slices)	70 °C/3 min at 20:1 (water: potato, w/w)	– Rinsing for 1 min before blanching; – Use of citric acid (0.05 M) or acetic acid (0.15 M) in blanching water	Frying in palm oil (175 °C/4 min)	293 (for both treatments)	49% (for both treatments)	Kita et al. (2004)
Asterix (1 mm slices)	20 °C/60 min (soaking) at 20:1 (water: potato, w/w)	– Rinsing for 1 min before blanching; – Use of acetic acid (0.15 M) in blanching water	Frying in palm oil (175 °C/4 min)	60	90%	Kita et al. (2004)
Tivoli (2.2 × 37 mm slices, t × d)	70 °C/40 min at 67:1 (water: potato, w/w)	Rinsing for 1 min before blanching	Frying (190 °C/3.5 min) until moisture of ~ 1.7%	116	~ 98%	Pedreschi et al. (2004)
Tivoli (2.2 × 37 mm slices, t × d)	50 °C/70 min at 67:1 (water: potato, w/w)	Rinsing for 1 min before blanching	Frying (190 °C/3.5 min) until moisture of ~ 1.7%	28	~ 99%	Pedreschi et al. (2004)

n.p. Not performed

n.s.r. No significant reduction

n.a. Information or detailing not available

[a]Compared to a 1-step conventional blanching (with no combined methods)

fried products (Krokida et al. 2001; Tajner-Czopek et al. 2008). It must be emphasized that the blanching time is only one of the parameters directly linked to the efficiency of acrylamide and Maillard precursors leaching. In this sense, a higher leaching—which is a phenomenon derived from time, temperature, mass transfer resistances, and solute concentration—is probably the main responsible for the changes in color than just time itself.

Use of Combined Methods in Acrylamide Mitigation

Blanching is effective in diminishing acrylamide formation if enough temperature and time are applied in the process. Depending on the degree of leaching wanted, i.e., the final amount of acrylamide established as limit in the foodstuff, an industrial process may be compromised by time consuming. Using mild to high blanching temperatures (let us say 60–85 °C) may take 20, 30 min, or even more time of immersion for different processes, products, cultivars, or final cooking. To overcome this issue, there are several technological solutions to be combined to blanching towards to efficiency in acrylamide mitigation.

Soaking/Rinsing

As pointed above, blanching is only one of a great variety of methods used to reduce the amount of acrylamide in food. Some of these methods can be used in combination with blanching at different stages, depending on the characteristics of the process. To be specific, the alternative methods can be applied before, during, or after the immersion step. Choosing more than one method at different moments can also be an option. The simplest procedure that can be combined to blanching consists of a previous or posterior soaking or rinsing.

Soaking/rinsing as a preceding stage is intended to leach starch and cellular material directly adhered to the vegetable surface (Pedreschi et al. 2005; Shojaee-Aliabadi et al. 2013). A posterior soaking/rinsing is applied to leach out the acrylamide precursors present in the blanching water that surrounds the food piece. The latter situation may be a helpful alternative when there is not a sufficient renewal in the blanching water, for instance. In this case, a greater concentration of reducing sugars and asparagine should occur in the immersion solution, leading to a lower final leaching after the blanching step. The use of soaking can also be seen as a two-step leaching process, where a HTST blanching is usually performed first. This arrangement improved the results when blanching Verdi potato slices (85 °C/3.5 min) for chips making (Pedreschi et al. 2011). After a soaking step (50 °C/20 min) the chips presented 45% less acrylamide than the samples only blanched.

Additives

If adding another immersion step in fresh water represents a longer processing, using additives directly in the blanching water or in a secondary immersion step may substantially save time. In any case the protocols followed are the same as those involving blanching in fresh water. The only difference to the conventional blanching remains in the water bath, changed for an additive solution of known concentration. Although there is an extensive list of additives proven to be effective in acrylamide mitigation, the studies involving their association with blanching are limited to some organic acids, salts, and amino acids.

Acids

Lowering the blanching water pH with different organic acids represents a cheap and effective way to reduce acrylamide content. The mechanism of acrylamide mitigation using acids is involved with the blockage of the nucleophilic addition of the α-amino group of asparagine molecules to the carbonyl group of a dicarbonyl (e.g., fructose or glucose) in the Maillard reaction chain (Jung et al. 2003). This prevents the formation of a Schiff base, a precursor of acrylamide and Maillard compounds (Vinci et al. 2012). The conversion of free nonprotonated amines ($-NH_2$) to protonated amines ($-NH_3^+$) can be achieved by providing hydrogen ions (H^+) through the dissociation of acids in an aqueous solution, in this case, the blanching solution.

Although there is a plenty of studies involving the use of acid solutions (dipping) to reduce acrylamide formation in foods, data with respect to processes where heat is applied along or after the immersion step are still scarce. The studies found, where acids are used in a blanching bath or on a secondary immersion step, report the application of citric, lactic and acetic acid to mitigate acrylamide formation on potato products (Kita et al. 2004; Mestdagh et al. 2008a; Vinci et al. 2011). When both methods are combined, the acrylamide formation is further reduced (compared to their single use). A higher acrylamide reduction is achieved for more concentrated acid solutions, allowing even a full mitigation in some cases. It is not appropriate, however, to use an acidic solution in high concentrations. In French fries processing, Vinci et al. (2011) verified that using citric acid and acetic acid in an industrial process resulted in a high percentage of French fries samples not in specification. As the use of acids may mischaracterize the original sensorial attributes of French fries, their use should be reduced to ensure minimum changes in flavor, aroma, texture, and color. Although at lab scale is seems that a great reduction in acrylamide content can be achieved at pH around 4.7 (Mestdagh et al. 2008a), much lower pH values may be needed to achieve a significant reduction of acrylamide in industrial scale processed potatoes (Vinci et al. 2012) or in processes involving short time blanching (Kita et al. 2004). It is possible that using acids could be an option for producing acid-flavored crisps (as sour cream and vinegar flavors) or naturally acid foodstuffs if few off-flavors are produced, even though no data or document was found to support this statement.

Salts

It is relevant to say that the conventional French fries and potato chips industrial processes usually apply a blanching step and a posterior sodium acid pyrophosphate (SAPP, inorganic salt) dipping. The application of SAPP is already an acidification step and provides a mild acidity to the blanching or dipping water. Actually, SAPP application is intended to avoid the darkening of blanched potatoes: during the thermal processing a ferrous-chlorogenic acid complex is formed, which oxidizes to ferri-dichlorogenic acid (a bluish-gray compound) when exposed to air (Camire et al. 2009; Vinci et al. 2012). The works of Fiselier et al. (2005), Mestdagh et al. (2008a) and Vinci et al. (2011) present relevant data in assessing the combination of blanching with this chemical in acrylamide mitigation, but further investigation is still needed. In the research performed by Mestdagh et al. (2008a) for instance, potato chips blanched in a 50 mM SAPP solution (pH 4.7) presented an 83% reduction in acrylamide content, when compared with only blanched samples. However, this concentration may represent 10–20 folds higher than that applied in industrial potato chips processing (Mestdagh et al. 2008a; Gupta 2009). A similar trend was found in the work of Truong et al. (2014), who used a two-step blanching process with a 0.5% SAPP solution (pH 4.4) in the second stage (Table 5.1), resulting in 86% less acrylamide in the final product. Using basic solutions is also proven to be an effective way to reduce acrylamide formation in potato crisps, as shown by Kita et al. (2004), who achieved a 74% reduction in acrylamide formation after soaking potato slices in a 1% NaOH solution. However, the use of basic solutions influences the overall appearance, taste, and flavor of the potato products, making them unacceptable from a sensory viewpoint.

The presence of different cations in the immersion solution, as Na^+, K^+, Ca^{2+}, or Fe^{3+}, also plays an important role in acrylamide mitigation (Pedreschi et al. 2010). These cations can also interact with asparagine, preventing the formation of Schiff bases. The higher the concentration of cations in the dipping solution, the higher the prevention of acrylamide formation. Besides, polyvalent cations are capable of providing a more pronounced mitigation of acrylamide intermediates (Gökmen and Şenyuva 2007).

In a model system, Mestdagh et al. (2008c) found no influence of adding NaCl directly to a potato mix (200 μmol g^{-1}, no blanching) in the acrylamide formation of the final product. However, the same authors verified that applying an immersion blanching containing 0.05 and 0.1 mol L^{-1} of NaCl for potato slices could result in a reduction of 28% and 43% in the acrylamide content of the crisps, respectively. For a two-step process the acrylamide mitigation is, consequently, further improved. Pedreschi et al. (2010) for example studied the use of blanching and a posterior soaking in a 1% NaCl solution as a mean of controlling acrylamide formation in potato chips (Table 5.2). The authors found in this process a 62%

reduction in the acrylamide content of the final product, compared to that obtained only by blanching. In another case, it was possible to reduce in 47% and 89% the acrylamide formation in potato chips using 0.5% sodium bisulfite and 0.3% calcium chloride solutions as blanching media, respectively (Ou et al. 2008). The use of salts, especially NaCl, must be controlled to avoid producing too salty foods.

Like most food systems, the addition of salts also affects the sensory properties of blanched vegetables. Ou et al. (2008) for example found that the addition of 5% $CaCl_2$ in the blanching water was capable of reducing the final acrylamide content of potato chips to undetectable amounts, while a two folds higher brittleness was verified (1500 g breaking force), compared to samples subjected to a blank, a cysteine or a $NaHSO_3$ immersion. This difference was considered positive according to the authors, but there are no discussions regarding the taste. Complementarily, using approximately ten folds less $CaCl_2$ in the blanching water (0.05 mol L^{-1} or $\sim 0.55\%$), Mestdagh et al. (2008a) verified a 64% reduction in the acrylamide content of potato chips (compared to a fresh water-blanching), an improved texture (crispness and snap), and taste and color comparable to a blank sample (with no additive). In this case, a higher bitterness was identified by the panelists, which was statistically equal to the blank. These same authors also verified that using $\sim 0.6\%$ NaCl in the blanching water gave 28% less acrylamide in potato chips (compared to a fresh water-blanching) and a perceptible salt taste. Overall taste, color, and texture were not different from blank samples.

Amino Acids

The use of amino acids in combination with blanching was also verified. The main possible mechanisms involved in this method of acrylamide mitigation can either be a competition with asparagine in the Maillard reactions and/or a covalent binding between the amino compounds and acrylamide by a Michael-type addition, producing a corresponding 3-(alkylamino)propionamide (Zamora et al. 2010). In this type of treatment, it is common to use amino acids such as glycine, L-lysine, and cysteine. However, the use of one or a mixture of these compounds also affects the sensory properties of the food products. The higher the amino acid content, the higher the formation of darker products. This happens due to the extra reactive amino acids present in the media, capable of forming more Maillard compounds. In this sense, care should be taken to avoid applying a concentration too high of amino acids in any food subjected to Maillard reactions. Besides, depending on the amount and type of amino acids used, off-flavors may appear and depreciate the food quality, even in combination with blanching (Mestdagh et al. 2008a; Morales et al. 2008).

Fermentation

An interesting method to combine with potato processing is described by Baardseth et al. (2006) in a study where blanching is followed by a posterior lactic fermentation of potato rods in French fries production (Table 5.1). According to the authors, the use of *L. plantarum* NC8 reduced glucose and fructose contents in 98% and 83% after 1 h incubation, respectively. Consequently, after blanching and 45 min or 2 h of fermentation (37 °C), the acrylamide content of the samples decreased by 79% and 94%, respectively. No differences were found in asparagine content after 5 h of fermentation and a low to mid reduction was verified in other amino acids as alanine, arginine, phenylalanine, and serine. With these results, the authors concluded that the reduced sugar content was the factor responsible for the acrylamide mitigation. However, as in the fermentation process lactic acid was also produced, the pH of the incubation media (not measured) was also lowered, contributing to the acrylamide reduction. Also, for the unfermented samples there were not immersions in pure water after blanching (i.e., a 45 min or 2 h immersion at 37 °C in fresh water) simulating real blank samples. Regarding this, leaching is probably another factor responsible for part of the acrylamide mitigation effect, which was not taken into account. These facts do not diminish the merit of the findings, of course, but they should support the discussion around the data found.

The use of sole or combined blanching and fermentation also revealed different color formation in French fries after frying, as can be seen in Fig. 5.2. Both blanching and fermentation processes were effective in producing lighter French fries. The higher the fermentation period, the lighter the French fries. When combined, the methods were further effective in producing lighter potatoes. Moreover, the color formation followed the acrylamide content of the samples to some extent. However, when different cooking temperatures are used, the color cannot be used as a reliable predictor of acrylamide concentration (Taubert et al. 2004).

Even without a sensory assessment, the authors performed a limited consumer test where no differences in taste were detected between the fermented fries and the unfermented control. It is also important to emphasize that French fries usually present a golden color, while crisps are usually lighter. This said, if some Maillard browning is desirable in French fries or any other product, it will inevitable come with the cost of some acrylamide formation.

Asparaginase

Since not only reducing sugars but also the asparagine content is an important factor in acrylamide production, the use of specific enzymes in asparagine hydrolysis emerged as an alternative process involved in the lowering of acrylamide formation. These enzymes are called asparaginases (L-asparagine amidohydrolase) and catalyze the hydrolysis of asparagine into molecules of ammonia and aspartic acid. Nowadays there are two commercial asparaginase products available and focused in the food industry: PreventASe® (DSM, The Netherlands) and Acrylaway®

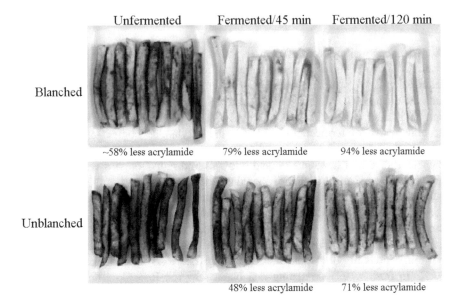

Fig. 5.2 French fries (var. Beate, 9 × 9 mm rods) subjected to blanching and/or fermentation followed by frying. Blanching at 80 °C for 5 min, fermentation with *L. plantarum* NC8 at 37 °C and two steps frying: the 1st at 170 °C for 3 min and the 2nd at 170 °C for 2 min 15 s after cooling in air for 5 min (Baardseth et al. 2006, modified)

(Novozymes A/S, Denmark). The former is obtained from a recombinant *A. niger* strain, showing improved activity in pH 4–5 at 50 °C. The latter is produced by *A. oryzae* and present higher activity in neutral pH at 60 °C (Pedreschi et al. 2008; Ciesarová 2016; Xu et al. 2016). The use of one or another enzyme type and its quantity will depend on the product and process characteristics.

Depending on the concentration of asparaginase used, it can be a quick alternative method to be applied in combination with blanching. Zuo et al. (2015) studied the acrylamide formation on French fries (Table 5.1) and verified that using asparaginase in the blanching water (10 U of L-asparaginase per mL) results in potatoes with 83% less acrylamide for a 8 min blanching, while a 39% reduction was achieved using only pure water for 10 min of immersion (both at 80 °C). Similarly, Pedreschi et al. (2008) also found 39% less acrylamide in French fries after a regular blanching (75 °C/10 min, Table 5.1). Using a posterior step of soaking in an asparaginase solution (10 ASNU per mL, 40 °C/20 min) it was possible to improve acrylamide reduction to 77%. With the existence of an extra soaking step and a longer time for enzyme-substrate contact in the work of Pedreschi and coworkers, one could imagine a higher acrylamide mitigation should be attained, compared with that verified by Zuo et al. (2015). One possible explanation for this slightly lower acrylamide reduction could be the different temperatures applied in the two processes, which must influence the enzyme activity distinctively.

Ciesarová (2016) suggests that blanching and asparaginase application may be not only additive methods but they could also play some kind of synergistic effect in asparagine removal. The author explains that some microstructural changes in the potato tissue may occur during blanching, promoting the L-asparagine diffusion from inside the vegetable to the enzyme-containing solution. Although this hypothesis has not yet been elucidated, this effect can be seen in the results of three trials made by Pedreschi et al. (2011). Let us call these trials: (a) soaking in asparaginase solution (50 °C/20 min); (b) blanching (85 °C/3.5 min) + soaking in water (50 °C/20 min); and (c) blanching + soaking in asparaginase solution. For the trials a, b and c, the acrylamide reduction was 15%, 62%, and 92%, respectively. It means that the single effect of using an asparaginase soaking (a) was doubled to 30% when used after a blanching process (c).

According to manufacturers and authors from different studies, the use of asparaginase in foods has no influence on taste, appearance, or any other sensory characteristic of the product (Morales et al. 2008; Hendriksen et al. 2009). However, after processing it will consequently affect the Maillard reactions, leading to a less pronounced flavor and color. The use of asparaginase restricts the process to a certain range of pH and temperature that should be used to avoid long immersion periods and/or asparaginase denaturation. Besides, the acquisition of asparaginase is costly (as usually enzymes are) and may represent a limitation if the consumer cannot absorb the price embedded in the technology, as in the case of products of low added value.

Processing Conditions

Even if the blanching process is optimized to reduce acrylamide formation, there are other parameters which will influence the acrylamide formation. As most of them are unrelated to the blanching step, some will be briefly described to give a better comprehension of the whole scenario involving the acrylamide occurrence in blanched foods. The main parameters to be controlled are those involving the cooking process: time, temperature, and method of cooking. For higher temperatures, higher total acrylamide contents were found in blanched Agria potato chips fried at 170 °C/5 min, 180 °C/4.15 min and 190 °C/3.5 min (Shojaee-Aliabadi et al. 2013). Increasing the temperature from 170 to 190 °C resulted in 60% more acrylamide in the fries. The times of thermal processing used were chosen to achieve a final 1.5% moisture. It is certain that thermal processes as frying at the range of approximately 120–190 °C have a tendency of producing more acrylamide at higher temperatures. However, the use of temperatures higher than 200 °C and prolonged heating are methods that may also result in less acrylamide in some cases (Taubert et al. 2004; Gökmen and Şenyuva 2006), not due to a lower formation but to acrylamide degradation instead.

The cooking method will determine the product final characteristics and the acrylamide formation will be largely reflected by it. In the work of Skog et al.

(2008), potato wedges were blanched in boiling water prior to roasting (4 min boiling) and frying (20 min boiling). Even using a longer blanching time, the fried potatoes (unknown temperature, 4 min frying) presented 3–6 folds the acrylamide content of the roasted potatoes (225 °C/25 min, wedges turned over after 10 min). This result could be explained by the higher thermal energy transfer that occur in deep-frying, compared to regular baking. Additively, the whole vegetable piece suffers dehydration in deep-frying while in an oven there is always a surface limited to the tray and other exposed to hot air. The low moisture is crucial to develop Maillard products and acrylamide.

Regarding the blanching step, it will lose effectiveness in acrylamide mitigation if the process design is incorrect. In a continuous or semi-continuous process for example, the water or solution of immersion needs to be renewed to avoid getting too loaded with amino acids, sugars, and other components from the vegetable tissue. The higher the concentration of soluble components in the blanching solution, the lower will be the removal of components from the vegetable. Mestdagh et al. (2008b) for example verified that reusing the blanching water in a potato fries process resulted in 10% lower reducing sugars removal and 10% lower reduction in the final acrylamide content of the final product. Using higher water:vegetable ratios and a proper water renewal may be suitable practices in the improvement of leaching efficiency. These actions will cost energy, water and a higher post-treatment effort though.

The vegetable and cultivar used will also exert influence on the final acrylamide formation. On investigating the acrylamide content of potato chips, Shojaee-Aliabadi et al. (2013) verified a remarkably higher acrylamide formation in unblanched Sante potatoes, followed by Agria variety and finally the Savalan type. Although the Savalan cultivar presented 47% and 67% more asparagine than the Agria and Sante varieties, it presented 23% and 54% less reducing sugars than the respectives. In this case, it seems that the amount of reducing sugars played the main role on limiting acrylamide formation. The use of blanching did not change the trend followed by the unblanched samples. In fact, in some cases the application of a blanching process may affect similarly different potato cultivars, with respect to the percentage of acrylamide precursor's removal. Vinci et al. (2010) found that after blanching 10 different potato cultivars (Table 5.1) the removal of reducing sugars fluctuated near 50% for all varieties. Following this trend in reducing sugars removal, the acrylamide reduction of nine varieties ranged in 48–66%, while for the Victoria cultivar a reduction of 86% was achieved.

Another technology studied as a treatment done after blanching and before frying was the pre-drying (Pedreschi et al. 2007). In this study, blanched potatoes (85 °C/3.5 min, Desirée cultivar) were subjected to air drying (60 °C and air velocity of 1 m/s) and different frying temperatures (120, 150, and 180 °C until 1.8% moisture content). The combination of blanching and pre-drying resulted in approximately 900% and 15% more acrylamide after frying at 120 and 150 °C, while ∼56% less acrylamide was verified when frying at 180 °C. For the non-dried blanched samples the acrylamide formation was ∼33–52% lower for the three temperatures of cooking. The application of a pre-drying is justified by obtaining improved color, improved texture and reduced oil uptake at the cost of more acrylamide formation than only blanched samples. Complementarily, care must be taken when evaluating the acrylamide reduction or increase as a fraction (in percentage) since the total acrylamide content and total acrylamide consumed is what will indeed represent or not a health issue.

The time and temperature of storage before and after processing may have influence on acrylamide precursors and its final content. At lower temperatures (above the freezing point), potatoes presented higher reducing sugar contents and, consequently, higher final acrylamide content. Additionally, the higher the storage period, the higher the reducing sugars content (Fiselier et al. 2005; Cummins et al. 2008).

Although a dextrose dipping is a common step applied in fried potatoes industries, studies involving the use of blanching as a method of acrylamide mitigation in a process line containing a dextrose dip step are still scarce (Vinci et al. 2011, 2012). Dextrose is used to improve and standardize the color of the final product (golden color) and will affect acrylamide formation in potato products since it is a reducing sugar (Taeymans et al. 2004).

Final Considerations

Blanching is a method that must be considered in the acrylamide mitigation of low processed vegetables due to its simplicity, relatively low cost and multipurpose characteristics (e.g., enzyme inactivation, microbial reduction, acrylamide precursors leaching, color stabilizing). When thinking of an industrial line producing French fries, the use of blanching is not a choice: it must be part of the process; otherwise the potatoes will get a burnt flavor and a dark color due to the excess sugars. The blanching must be tuned properly in order to guarantee an optimized final set of quality attributes. Another process that should be taken into account in French fries processing is the par-frying associated to posterior freezing, which confer desirable textural properties to the final product. The application of these methods in a blanching-containing line was not considered in all researches where acrylamide content was assessed.

After taking into account the final attributes, using blanching temperatures of approximately 70–80 °C should improve the overall quality—lower acrylamide content, lower oil uptake, good color and texture, and optimized acceptance—of fried potatoes. Using higher blanching temperatures may improve efficiency but at the expense of inferior sensory characteristics. A recommended strategy to improve efficiency and final quality is to combine different methods and technologies with blanching in acrylamide mitigation. As every single method present advantages and disadvantages related to cost, time and quality, future studies should focus in combining more than one method in acrylamide reduction. The interaction of methods is likely to require less severe processing conditions and result in products presenting superior characteristics. The efficacy of the acrylamide mitigation is also dependent of an optimized process for a specific product. Finally, the use of such technologies and optimizations must be justified by a product accepted by consumers, both for its cost and its sensory characteristics.

References

Adams A, Hamdani S, Van Lancker F et al (2010) Stability of acrylamide in model systems and its reactivity with selected nucleophiles. Food Res Int 43:1517–1522. Doi:10.1016/j.foodres.2010. 04.033

Agblor A, Scanlon MG (2000) Processing conditions influencing the physical properties of French fried potatoes. Potato Res 43:163–177. Doi:10.1007/BF02357957

Anese M, Suman M, Nicoli MC (2010) Acrylamide removal from heated foods. Food Chem 119:791–794. Doi:10.1016/j.foodchem.2009.06.043

Baardseth P, Blom H, Skrede G et al (2006) Lactic acid fermentation reduces acrylamide formation and other Maillard reactions in French fries. J Food Sci 71:C28–C33. Doi:10.1111/j. 1365-2621.2006.tb12384.x

Banchero M, Pellegrino G, Manna L (2013) Supercritical fluid extraction as a potential mitigation strategy for the reduction of acrylamide level in coffee. J Food Eng 115:292–297. Doi:10.1016/ j.jfoodeng.2012.10.045

Belitz H-D, Grosch W, Schieberle P (2009) Amino acids, peptides, proteins. Food chemistry. Springer, Berlin, pp 8–89

Bhagat J, Kaur A, Chadha BS (2016) Single step purification of asparaginase from endophytic bacteria *Pseudomonas oryzihabitans* exhibiting high potential to reduce acrylamide in processed potato chips. Food Bioprod Process 99:222–230. Doi:10.1016/j.fbp.2016.05.010

Brathen E, Kita A, Knutsen SH, Wicklund T (2005) Addition of glycine reduces the content of acrylamide in cereal and potato products. J Agric Food Chem 53:3259–3264. Doi:10.1021/ jf048082o

Burch RS, Trzesicka A, Clarke M et al (2008) The effects of low-temperature potato storage and washing and soaking pre-treatments on the acrylamide content of French fries. J Sci Food Agric 88:989–995. Doi:10.1002/jsfa.3179

Camire ME, Kubow S, Donnelly DJ (2009) Potatoes and human health. Crit Rev Food Sci Nutr 49:823–840. Doi:10.1080/10408390903041996

Capuano E (2016) Lipid oxidation promotes acrylamide formation in fat-rich systems. In: Gökmen V (ed) Acrylamide in food. Academic Press, pp 309–324

Casado FJ, Sánchez AH, Montaño A (2010) Reduction of acrylamide content of ripe olives by selected additives. Food Chem 119:161–166. Doi:10.1016/j.foodchem.2009.06.009

Cha M (2013) Enzymatic control of the acrylamide level in coffee. Eur Food Res Technol 236:567–571. Doi:10.1007/s00217-013-1927-8

Chen H, Gu Z (2014) Effect of ascorbic acid on the properties of ammonia caramel colorant additives and acrylamide formation. J Food Sci 79:C1678–C1682. Doi:10.1111/1750-3841. 12560

Cheng K-W, Zeng X, Tang YS et al (2009) Inhibitory mechanism of naringenin against carcinogenic acrylamide formation and nonenzymatic browning in Maillard model reactions. Chem Res Toxicol 22:1483–1489. Doi:10.1021/tx9001644

Ciesarová Z (2016) Impact of l-asparaginase on acrylamide content in fried potato and bakery products. In: Gökmen V (ed) Acrylamide in food. Academic Press, pp 405–421

Cummins E, Butler F, Gormley R, Brunton N (2008) A methodology for evaluating the formation and human exposure to acrylamide through fried potato crisps. LWT - Food Sci Technol 41:854–867. Doi:10.1016/j.lwt.2007.05.022

Fiselier K, Hartmann A, Fiscalini A, Grob K (2005) Higher acrylamide contents in French fries prepared from "fresh" prefabricates. Eur Food Res Technol 221:376–381. Doi:10.1007/s00217-005-1183-7

Friedman M (2015) Acrylamide: inhibition of formation in processed food and mitigation of toxicity in cells, animals, and humans. Food Funct 6:1752–1772. Doi:10.1039/C5FO00320B

Friedman MA, Dulak LH, Stedham MA (1995) A lifetime oncogenicity study in rats with acrylamide. Fundam Appl Toxicol 27:95–105. Doi:10.1006/faat.1995.1112

Gökmen V (ed) (2016) Acrylamide in food: analysis, content and potential health effects. Academic Press, London

Gökmen V, Şenyuva HZ (2006) Study of colour and acrylamide formation in coffee, wheat flour and potato chips during heating. Food Chem 99:238–243. Doi:10.1016/j.foodchem.2005.06.054

Gökmen V, Şenyuva HZ (2007) Effects of some cations on the formation of acrylamide and furfurals in glucose-asparagine model system. Eur Food Res Technol 225:815–820. Doi:10.1007/s00217-006-0486-7

Gökmen V, Kocadağlı T, Göncüoğlu N, Mogol BA (2012) Model studies on the role of 5-hydroxymethyl-2-furfural in acrylamide formation from asparagine. Food Chem 132:168–174. Doi:10.1016/j.foodchem.2011.10.048

Granvogl M, Jezussek M, Koehler P, Schieberle P (2004) Quantitation of 3-aminopropionamide in potatoes—a minor but potent precursor in acrylamide formation. J Agric Food Chem 52:4751–4757. Doi:10.1021/jf049581s

Gupta MK (2009) Industrial frying. In: Sahin S, Sumnu SG (eds) Advances in deep-fat frying of foods. CRC Press, Boca Raton, pp 263–287

Hamzalıoğlu A, Gökmen V (2012) Role of bioactive carbonyl compounds on the conversion of asparagine into acrylamide during heating. Eur Food Res Technol 235:1093–1099. Doi:10.1007/s00217-012-1839-z

Hedegaard RV, Granby K, Frandsen H et al (2008) Acrylamide in bread. Effect of prooxidants and antioxidants. Eur Food Res Technol 227:519–525. Doi:10.1007/s00217-007-0750-5

Hendriksen HV, Kornbrust BA, Ostergaard PR, Stringer MA (2009) Evaluating the potential for enzymatic acrylamide mitigation in a range of food products using an asparaginase from *Aspergillus oryzae*. J Agric Food Chem 57:4168–4176. Doi:10.1021/jf900174q

Hidalgo FJ, Delgado RM, Zamora R (2011) Positive interaction between amino and sulfhydryl groups for acrylamide removal. Food Res Int 44:1083–1087. Doi:10.1016/j.foodres.2011.03.013

Hogervorst JGF, Baars B-J, Schouten LJ et al (2010) The carcinogenicity of dietary acrylamide intake: a comparative discussion of epidemiological and experimental animal research. Crit Rev Toxicol 40:485–512. Doi:10.3109/10408440903524254

Johnson KA, Gorzinski SJ, Bodner KM et al (1986) Chronic toxicity and oncogenicity study on acrylamide incorporated in the drinking water of Fischer 344 rats. Toxicol Appl Pharmacol 85:154–168. Doi:10.1016/0041-008X(86)90109-2

Jung MY, Choi DS, Ju JW (2003) A novel technique for limitation of acrylamide formation in fried and baked corn chips and in French fries. J Food Sci 68:1287–1290. Doi:10.1111/j.1365-2621.2003.tb09641.x

Kita A, Bråthen E, Knutsen SH, Wicklund T (2004) Effective ways of decreasing acrylamide content in potato crisps during processing. J Agric Food Chem 52:7011–7016. Doi:10.1021/jf049269i

Kotsiou K, Tasioula-Margari M, Kukurová K, Ciesarová Z (2010) Impact of oregano and virgin olive oil phenolic compounds on acrylamide content in a model system and fresh potatoes. Food Chem 123:1149–1155. Doi:10.1016/j.foodchem.2010.05.078

Krokida MK, Oreopoulou V, Maroulis ZB, Marinos-Kouris D (2001) Effect of pre-drying on quality of French fries. J Food Eng 49:347–354. Doi:10.1016/S0260-8774(00)00233-8

Li D, Chen Y, Zhang Y et al (2012) Study on mitigation of acrylamide formation in cookies by 5 antioxidants. J Food Sci 77:C1144–C1149. Doi:10.1111/j.1750-3841.2012.02949.x

Mariotti M, Cortés P, Fromberg A et al (2015) Heat toxicant contaminant mitigation in potato chips. LWT - Food Sci Technol 60:860–866. Doi:10.1016/j.lwt.2014.09.023

Mesias M, Morales FJ (2016) Acrylamide in bakery products. In: Gökmen V (ed) Acrylamide in food. Academic Press, pp 131–157

Mestdagh F, De Wilde T, Delporte K et al (2008a) Impact of chemical pre-treatments on the acrylamide formation and sensorial quality of potato crisps. Food Chem 106:914–922. Doi:10.1016/j.foodchem.2007.07.001

Mestdagh F, De Wilde T, Fraselle S et al (2008b) Optimization of the blanching process to reduce acrylamide in fried potatoes. LWT - Food Sci Technol 41:1648–1654. Doi:10.1016/j.lwt.2007.10.007

Mestdagh F, Maertens J, Cucu T et al (2008c) Impact of additives to lower the formation of acrylamide in a potato model system through pH reduction and other mechanisms. Food Chem 107:26–31. Doi:10.1016/j.foodchem.2007.07.013

Morales F, Capuano E, Fogliano V (2008) Mitigation strategies to reduce acrylamide formation in fried potato products. Ann N Y Acad Sci 1126:89–100. Doi:10.1196/annals.1433.051

Mustafa A, Andersson R, HellEnäs K-E et al (2008) Moisture enhances acrylamide reduction during storage in model studies of rye crispbread. J Agric Food Chem 56:11234–11237. Doi:10.1021/jf801405q

Ou S, Lin Q, Zhang Y et al (2008) Reduction of acrylamide formation by selected agents in fried potato crisps on industrial scale. Innov Food Sci Emerg Technol 9:116–121. Doi:10.1016/j.ifset.2007.06.008

Parker JK, Balagiannis DP, Higley J et al (2012) Kinetic model for the formation of acrylamide during the finish-frying of commercial French fries. J Agric Food Chem 60:9321–9331. Doi:10.1021/jf302415n

Pedreschi F, Kaack K, Granby K (2004) Reduction of acrylamide formation in potato slices during frying. LWT - Food Sci Technol 37:679–685. Doi:10.1016/j.lwt.2004.03.001

Pedreschi F, Moyano P, Kaack K, Granby K (2005) Color changes and acrylamide formation in fried potato slices. Food Res Int 38:1–9. Doi:10.1016/j.foodres.2004.07.002

Pedreschi F, Kaack K, Granby K (2006) Acrylamide content and color development in fried potato strips. Food Res Int 39:40–46. Doi:10.1016/j.foodres.2005.06.001

Pedreschi F, León J, Mery D et al (2007) Color development and acrylamide content of pre-dried potato chips. J Food Eng 79:786–793. Doi:10.1016/j.jfoodeng.2006.03.001

Pedreschi F, Kaack K, Granby K (2008) The effect of asparaginase on acrylamide formation in French fries. Food Chem 109:386–392. Doi:10.1016/j.foodchem.2007.12.057

Pedreschi F, Granby K, Risum J (2010) Acrylamide mitigation in potato chips by using NaCl. Food Bioprocess Technol 3:917–921. Doi:10.1007/s11947-010-0349-x

Pedreschi F, Mariotti S, Granby K, Risum J (2011) Acrylamide reduction in potato chips by using commercial asparaginase in combination with conventional blanching. LWT - Food Sci Technol 44:1473–1476. Doi:10.1016/j.lwt.2011.02.004

Rivas-Jimenez L, Ramírez-Ortiz K, González-Córdova AF et al (2016) Evaluation of acrylamide-removing properties of two *Lactobacillus* strains under simulated gastrointestinal conditions using a dynamic system. Microbiol Res 190:19–26. Doi:10.1016/j.micres.2016.04.016

Rydberg P, Eriksson S, Tareke E et al (2003) Investigations of factors that influence the acrylamide content of heated foodstuffs. J Agric Food Chem 51:7012–7018. Doi:10.1021/jf034649+

Sanghvi G, Bhimani K, Vaishnav D et al (2016) Mitigation of acrylamide by l-asparaginase from *Bacillus subtilis* KDPS1 and analysis of degradation products by HPLC and HPTLC. Springerplus 5:533. Doi:10.1186/s40064-016-2159-8

Shakambari G, Birendranarayan AK, Angelaa Lincy MJ et al (2016) Hemocompatible glutaminase free l-asparaginase from marine *Bacillus tequilensis* PV9W with anticancer potential modulating p53 expression. RSC Adv 6:25943–25951. Doi:10.1039/C6RA00727A

Shojaee-Aliabadi S, Nikoopour H, Kobarfard F et al (2013) Acrylamide reduction in potato chips by selection of potato variety grown in Iran and processing conditions. J Sci Food Agric 93:2556–2561. Doi:10.1002/jsfa.6076

Skog K, Viklund G, Olsson K, Sjöholm I (2008) Acrylamide in home-prepared roasted potatoes. Mol Nutr Food Res 52:307–312. Doi:10.1002/mnfr.200700240

Sun Z, Li D, Liu P et al (2016) A novel l-asparaginase from *Aquabacterium* sp. A7-Y with self-cleavage activation. Antonie Van Leeuwenhoek 109:121–130. Doi:10.1007/s10482-015-0614-0

Taeymans D, Wood J, Ashby P et al (2004) A review of acrylamide: an industry perspective on research, analysis, formation and control. Crit Rev Food Sci Nutr 44:323–347. Doi:10.1080/10408690490478082

Tajner-Czopek A, Figiel A, Carbonell-Barrachina ÁA (2008) Effects of potato strip size and pre-drying method on french fries quality. Eur Food Res Technol 227:757–766. Doi:10.1007/s00217-007-0784-8

Tareke E, Rydberg P, Karlsson P et al (2000) Acrylamide: a cooking carcinogen? Chem Res Toxicol 13:517–522. Doi:10.1021/tx9901938

Tareke E, Rydberg P, Karlsson P et al (2002) Analysis of acrylamide, a carcinogen formed in heated foodstuffs. J Agric Food Chem 50:4998–5006. Doi:10.1021/jf020302f

Taubert D, Harlfinger S, Henkes L et al (2004) Influence of processing parameters on acrylamide formation during frying of potatoes. J Agric Food Chem 52:2735–2739. Doi:10.1021/jf035417d

Törnqvist M (2005) Acrylamide in food: the discovery and its implications: a historical perspective. In: Friedman M, and Mottram D (eds) Chemistry and safety of acrylamide in food. Springer, 1–19. doi: 10.1007/0-387-24980-X_1

Truong VD, Pascua YT, Reynolds R et al (2014) Processing treatments for mitigating acrylamide formation in sweetpotato French fries. J Agric Food Chem 62:310–316. Doi:10.1021/jf404290v

Urbančič S, Kolar MH, Dimitrijević D et al (2014) Stabilisation of sunflower oil and reduction of acrylamide formation of potato with rosemary extract during deep-fat frying. Food Sci Technol 57:671–678. Doi:10.1016/j.lwt.2013.11.002

Viklund GÅI, Olsson KM, Sjöholm IM, Skog KI (2010) Acrylamide in crisps: effect of blanching studied on long-term stored potato clones. J Food Compos Anal 23:194–198. Doi:10.1016/j.jfca.2009.07.009

Vinci RM, Mestdagh F, De Muer N et al (2010) Effective quality control of incoming potatoes as an acrylamide mitigation strategy for the French fries industry. Food Addit Contam Part A Chem Anal Control Expo Risk Assess 27:417–425. Doi:10.1080/19440049.2011.639094

Vinci RM, Mestdagh F, Van Poucke C et al (2011) Implementation of acrylamide mitigation strategies on industrial production of French fries: challenges and pitfalls. J Agric Food Chem 59:898–906. Doi:10.1021/jf1042486

Vinci RM, Mestdagh F, De Meulenaer B (2012) Acrylamide formation in fried potato products— present and future, a critical review on mitigation strategies. Food Chem 133:1138–1154. Doi:10.1016/j.foodchem.2011.08.001

Xu X, An X (2016) Study on acrylamide inhibitory mechanism in Maillard model reaction: effect of p-coumaric acid. Food Res Int 84:9–17. Doi:10.1016/j.foodres.2016.03.020

Xu C, Yagiz Y, Marshall S et al (2015) Application of muscadine grape (*Vitis rotundifolia* Michx.) pomace extract to reduce carcinogenic acrylamide. Food Chem 182:200–208. Doi:10.1016/j.foodchem.2015.02.133

Xu F, Oruna-Concha MJ, Elmore JS (2016) The use of asparaginase to reduce acrylamide levels in cooked food. Food Chem 210:163–171. Doi:10.1016/j.foodchem.2016.04.105

Yaylayan VA, Stadler RH (2005) Acrylamide formation in food: a mechanistic perspective. J AOAC Int 88:262–267

Zamora R, Hidalgo FJ (2008) Contribution of lipid oxidation products to acrylamide formation in model systems. J Agric Food Chem 56:6075–6080. Doi:10.1021/jf073047d

Zamora R, Delgado RM, Hidalgo FJ (2010) Model reactions of acrylamide with selected amino compounds. J Agric Food Chem 58:1708–1713. Doi:10.1021/jf903378x

Zhang Y, Zhang Y (2008) Effect of natural antioxidants on kinetic behavior of acrylamide formation and elimination in low-moisture asparagine–glucose model system. J Food Eng 85:105–115. Doi:10.1016/j.jfoodeng.2007.07.013

Zuo S, Zhang T, Jiang B, Mu W (2015) Reduction of acrylamide level through blanching with treatment by an extremely thermostable l-asparaginase during French fries processing. Extremophiles 19:841–851. Doi:10.1007/s00792-015-0763-0

Zyzak DV, Sanders RA, Stojanovic M et al (2003) Acrylamide formation mechanism in heated foods. J Agric Food Chem 51:4782–4787. Doi:10.1021/jf034180i

Chapter 6
Impact of Blanching on the Performance of Subsequent Drying

Felipe Richter Reis

Abstract Drying is an ancient method designed to increase food shelf-life by means of water removal. The use of blanching prior to drying is a common practice, especially when low drying temperatures are used, because blanching inactivates enzymes responsible for undesirable sensory changes. However, a number of reports suggest that blanching is also able to improve the drying performance. In this chapter, such studies will be presented and discussed. Figures showing the positive impact of blanching on drying performance include data on drying kinetics, water activity kinetics, and effective moisture diffusivity, all of them showing that blanching is usually favorable for accelerating the subsequent drying process. Proposed mechanisms for this effect include cell membrane disruption, softening, removal of natural waxy layers, loss of turgor, and pores enlargement. Nevertheless, a couple of studies showing insignificant or negative effects of blanching on drying performance are also available. The impact of blanching on dried product quality is also dealt with in this chapter since product quality assessment is usually carried out in drying studies. Since not all foods behave the same way, the choice for blanching before drying must be made on the basis of experiments, thereby assuring that the impact of blanching will be favorable to the drying process and the product quality achieved will be desirable.

Keywords Blanching · Drying · Food · Drying kinetics

According to Mujumdar (2014), "drying commonly describes the process of thermally removing volatile substances (moisture) to yield a solid product." This reference work also states that the drying rate depends on the rate at which heat is transferred from the surroundings to the solid being dried, causing surface moisture to evaporate and the rate at which internal moisture is transferred to the solid surface to be evaporated due to heat transfer.

F. Richter Reis (✉)
Food Technician Course, Instituto Federal do Paraná, Campus Jacarezinho, Jacarezinho, Paraná, Brazil
e-mail: felipe.reis@ifpr.edu.br

© Springer International Publishing AG 2017
F. Richter Reis (ed.), *New Perspectives on Food Blanching*,
DOI 10.1007/978-3-319-48665-9_6

123

Various reports dealing with the effect of blanching pretreatments on the drying rate of miscellaneous foods were found in Scopus database in a search comprising the last ten years. Such results will be presented and discussed in this chapter. Usually, these studies also reported the results of quality assessments of the dried products. Nevertheless, only studies comprising at least the effect of blanching on drying performance will be presented here. Reports on the effect of blanching on dried product quality, solely, are dealt with in the other chapters. As done in Chap. 2, the studies will be presented chronologically and divided by product type.

In the middle of the last decade, Arévalo-Pinedo and Murr (2006) studied the vacuum drying kinetics of pumpkin as affected by blanching or freezing thawing pretreatments. They found that both pretreatments increased the drying rates, but freezing thawing was more effective than blanching in this sense. Since blanching was performed at high temperature (95 °C) and freezing was performed slowly, in a freezer at −20 °C, the hypotheses for explaining the observed behavior are the heat-induced cell damage during blanching and the cell disruption by sharp ice crystals during slow freezing, which eventually led to easier moisture removal during subsequent drying. The osmotic dehydration of pumpkin was studied by Kowalska et al. (2008), who found that a blanching pretreatment was effective for enhancing the decrease in water content as compared to unblanched pumpkin. In addition, the authors observed that the parameter called water loss presented higher values for blanched than for frozen samples. On the other hand, their results showed that blanching, such as freezing, undesirably increased the solids gain, being this effect less pronounced for the starch syrup osmotic solution, probably due to the low diffusivity of higher molecular weight substances contained in the syrup. With regard to effective diffusivities of sugars and moisture, they concluded that blanching resulted in increased values of these parameters when glucose solution was used. This behavior can be attributed to the small size of the glucose molecule thus facilitating diffusion and to the high concentration of the glucose solution (49.5%). The use of blanching before freeze-drying of pumpkin resulted in dried pumpkins of lower water content when compared to unblanched samples (Ciurzyńska et al. 2014). The mechanism behind this effect was probably cell disruption during blanching thus facilitating water removal during subsequent drying. The authors found no significant effect of blanching on color. Workneh et al. (2014) found that blanching of pumpkin slices prior to sun or oven drying resulted in significantly lower drying time and increased drying rates. Molina Filho et al. (2016) studied the air drying of pumpkin as preceded by blanching or pectin coating, finding that blanching resulted in higher evaporated water flux and faster drying, which was ascribed to thermal damage of the tissue and decrease in solids content.

Blasco et al. (2006) investigated the effect of blanching and hot air drying conditions on engineering aspects of turmeric drying, finding that blanching in boiling water for 4 min increased the values of effective moisture diffusivity and kinetic parameters of empirical models at all drying conditions. They affirmed that blanching affects the turmeric rhizome structure thus increasing the drying rate. Their results lead to the conclusion that the accelerating effect of blanching on drying at 60–100 °C is more pronounced at higher drying temperatures, which could be attributed to an additive effect of high drying temperature on the effect of blanching. Additionally, activation energy for moisture diffusion of blanched samples was lower than that of unblanched samples, suggesting that blanching facilitates moisture removal from turmeric.

Kadam et al. (2006) dried cauliflower in a solar drier and elucidated the effect of blanching combined with chemical pretreatments on drying time and quality of the dried product. They found that neither blanching nor chemical pretreatments affect the cauliflower drying time. On the other hand, product quality, as expressed in terms of rehydration capacity, ascorbic acid content, browning and sensory scores, was the best when a 3 min blanching in boiling water followed by a dipping in 1.0% sodium meta-bisulphate were used.

For potatoes, sometimes drying was proposed as substitute of frying in order to eliminate or reduce the oil content of potato snacks. Blanching was combined with drying and its impact on drying kinetics and product quality attributes was assessed in a couple of reports. For example, low pressure superheated steam-dried potatoes that were previously blanched at 90 °C for 5 min showed better retention of the original color and appropriate hardness, which were attributed to the inactivation of POD/leaching of reducing sugars and to starch gelatinization caused by blanching (Leeratanarak et al. 2006). They also found that blanched samples dried faster than unblanched ones, which was attributed to easier water removal during drying as a result of the softening effect of blanching. Another report showed that the combination of blanching and freezing enhanced the positive impact of blanching on the quality of low pressure superheated steam-dried potatoes, as denoted by higher lightness (L*), lower redness (a*), and higher yellowness (b*) along with a more preserved microstructure associated with a better texture (Pimpaporn et al. 2007). They also observed that the accelerating effect of blanching on the drying kinetics was enhanced by combining it with freezing, especially at lower drying temperatures (70 °C). When blanching was followed by various freezing/thawing cycles, low pressure superheated steam-dried potatoes presented higher hardness and toughness which were attributed to starch retrogradation during freezing (Kingcam et al. 2008). The use of various freezing/thawing cycles also enhanced the accelerating effect of blanching-and-freezing on drying rates (Fig. 6.1). Another study on potatoes led to the conclusion that blanching in boiling water for 2–5 min reduced the drying time for hot air drying without circulation, while for forced convection air drying and fluidized bed drying this effect was not observed (Hatamipour et al. 2007). The results might be explained by the fact that static air drying is significantly impacted by the effect of blanching, while the effect of moving air during forced convection air drying probably masks the effect of blanching. Additionally, they found that blanching preserved the color of cylinders of different varieties of potatoes, which may be attributed to the inhibition of PPO in the tissue.

Dried button and oyster mushrooms were pretreated by several techniques in order to elucidate their effect on the drying kinetics and on the values of effective moisture diffusivity (Walde et al. 2006). The authors found that blanching in 2% NaCl solution at 90 °C presented variable influence on the evaluated engineering

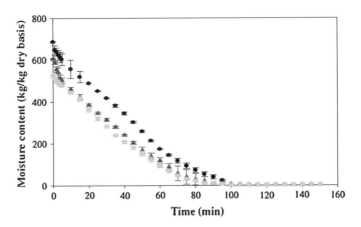

Fig. 6.1 Drying kinetics of potato chips with initial thickness of 3.5 mm, subjected to different pretreatments (*filled circle*) B + F, (*filled triangle*) B + F + T (3 cycles), (*filled square*) B + F + T (5 cycles) (B, Blanching, F, Freezing and T, Thawing) (Kingcam et al. 2008, modified)

parameters. Irrespective of the pretreatments used in that study, microwave drying was the fastest method while vacuum drying was the slowest. Interestingly, blanching augmented the microwave drying time. On the other hand, blanching decreased vacuum drying time which could be due to the damage caused by heat thus allowing freer moisture migration from the samples during drying. Fluidized bed drying was found to be the most suitable technique for yielding high quality dried mushrooms in a reasonable time, but the fluidized bed drying time was not affected by blanching. Cabinet drying time was not affected by blanching either. Immersion in acidified solutions, namely curd and fermented whey, decreased drying times and increased moisture diffusivities, which was attributed to the opening of the mushrooms pores by lactic acid bacteria.

The impact of non-chemical, mechanical pretreatments on the drying kinetics and quality of miscellaneous foods was reported by Yong et al. (2006). They found that blanching in water at 90 °C for 3 min was effective in reducing the drying time during heat-pump drying of chilies, especially when combined with drill holes. This effect was attributed to the destruction of cells and partial removal of the waxy layer of chilies, facilitating in this way the water removal during subsequent drying. Furthermore, using these techniques, the authors found that the color of chilies was well preserved. Results also showed that when freezing was combined with blanching and drill holes, there was an increase in drying rates, although there was also expressive color degradation.

Carrot was blanched or frozen before vacuum drying and the effect of these pretreatments on drying kinetics was evaluated (Arévalo-Pinedo and Murr 2007). The authors found that both pretreatments were suitable for accelerating the drying process, but freezing was more effective than blanching. They attributed this effect to cell disruption during slow freezing thus facilitating the removal of moisture during subsequent drying. In addition, the effect of blanching and freezing on

Fig. 6.2 Rehydration of blanched/ultrasound-assisted hot air-dried carrot. Visual aspect before (**a**) and after (**b**) rehydration (Soria et al. 2010)

drying kinetics was detectable only under low pressures (5 kPa), which might be due to the fact that the puffing effect that facilitates moisture removal of damaged cells was more pronounced under high vacuum. Another study on carrot showed that blanching increased drying rates and reduced drying time (Górnicki and Kaleta 2007). Such effect was higher for higher blanching times (6 min) and for pure water in detriment of 5% brine. Blanching was claimed to promote cell membrane disruption, loss of soluble solids and loss of turgor, mechanisms that probably facilitated water removal during drying. Soria et al. (2010) dried carrots by ultrasound-assisted convective drying preceded by blanching in boiling water for 1 min, finding that this pretreatment increased the drying rates and improved the rehydration of the dried product. In fact, the blanched rehydrated carrot presented a similar appearance to that of raw carrots (Fig. 6.2). The authors affirmed that the cellular network is loosened and the middle lamella is separated by blanching thus resulting in a softer carrot tissue that rehydrates better.

The osmotic dehydration of carrots was preceded by blanching in a study performed by Paredes Escobar et al. (2007). The authors found that longer blanching times led to higher dehydration rates (Fig. 6.3). Blanching was performed in water at 100 °C for times up to 30 s. This effect was attributed to cell damage, a hypothesis that was supported by calorimetric measurement of the samples thermal power. In addition, effective diffusivities of water and solute (sucrose) increased with an increase in blanching time, but this effect was lost during storage for up to 12 weeks (Fig. 6.4). The decrease in the effect of blanching on effective diffusivities was ascribed to the end of changes affecting the pore size of the cell wall after 12 weeks of storage.

In red chilies (*Capsicum annum* L.), blanching in hot water at 90 °C for 3 min before hot air drying at 60 °C, relative humidity of 20% and velocity of 0.5 m/s reduced drying times in 16.5%, which was attributed to heat-induced cell wall

Fig. 6.3 Variation of
moisture content with time
during osmotic dehydration at
40 °C (Paredes Escobar et al.
2007)

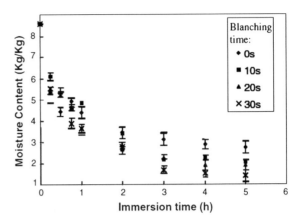

Fig. 6.4 Effect of blanching
pretreatments and long-term
storage on effective diffusion
coefficients of water
(Dew) and sucrose
(Des) during osmotic
dehydration of carrot
parenchyma (Paredes Escobar
et al. 2007)

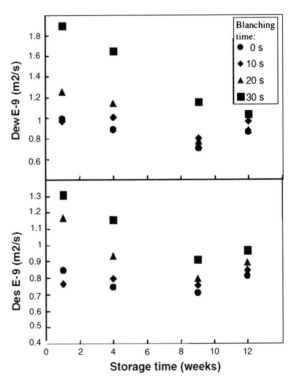

disruption (Hossain et al. 2007). Furthermore, color of chilies was more attractive after
blanching, which was attributed to increase in the relative amount of chili carotenoids
due to leaching of soluble solids. Vengaiah and Pandey (2007) dried sweet pepper
(*Capsicum annum* L.) with hot air at a temperature of 40–70 °C and velocity of
1.5 m/s after blanching in water at 95 °C for 5 min. They found that the drying times
were always lower for the blanched samples (Fig. 6.5). Tunde-Akintunde and Afolabi
(2009) found that drying of chili pepper (*Capsicum frutescens*) was accelerated by

Fig. 6.5 Drying of Sweet pepper at different temperature (Vengaiah and Pandey 2007)

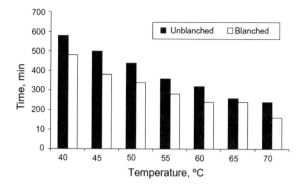

blanching in boiling water for 3 min. They used three methods of drying: sun, solar, and hot air, all three presenting lower process times when preceded by blanching. Such increase in drying rates was associated with higher values of effective moisture diffusivity. The authors hypothesized that the enlargement of chili pepper pores due to blanching reduced the resistance represented by the thick skin of peppers thus facilitating water removal by drying.

The drying of leek slices was affected by blanching, drying temperature, and product thickness (Doymaz 2008). Blanching increased the drying rates, as expected, which may be attributed to mechanisms discussed before in this chapter. In this sense, drying times of 570, 450 and 285 min for drying of 1 cm slices at 60, 70, and 80 were reduced to 495, 330, and 240 min when blanching was used. Effective moisture diffusivity also increased with blanching, which was attributed to the effect of blanching on the internal mass transfer of samples during drying. Rehydration was also superior in blanched leek slices, which can be attributed to easier water penetration in blanched cells.

Hot air drying of yacon (*Smallanthus sonchifolius*) was preceded by steam blanching (100 °C) for 4 min in a study by Scher et al. (2009). The authors found that blanching promoted faster decrease in moisture and water activity during drying in comparison to drying without blanching, as confirmed by drying curves and water activity curves (Fig. 6.6). More specifically, the time that blanched samples took to reach a constant value of a_w was shorter. Nevertheless, their results suggest that blanching reduces to some extent the concentrations of inulin, glucose and fructose. In this sense, steam blanching was found to be suitable to reduce drying time while promoting only mild nutrient losses. The effect of the blanching heat on cellular integrity probably facilitated the removal of moisture during drying, while nutrients were mostly preserved because steam blanching avoids their leaching.

Asparagus were blanched before drying in an attempt to increase the drying rates and improving product color (Bala et al. 2010). The authors found that blanching increased the drying rates for slices and longitudinal halves of asparagus, while for whole asparagus there was no effect. They hypothesized that tissues of split and sliced asparagus were possibly disrupted or loosened during blanching, ultimately

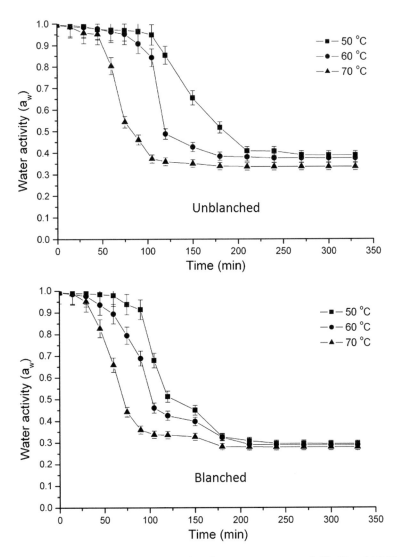

Fig. 6.6 Water activity as a function of drying time at temperatures of 50, 60 and 70 °C for non-blanched and blanched yacon (Scher et al. 2009, modified)

resulting in higher drying rates, while whole asparagus were not efficiently dried due to their thick skin. In addition, they reported that the effect of blanching was more intense for lower drying temperatures (50 °C) than for higher drying temperatures (70 °C). This behavior may be attributed to the high driving force at 70 °C either with or without blanching, while at 50 °C the driving force for drying is lower and therefore the impact of blanching is more significant.

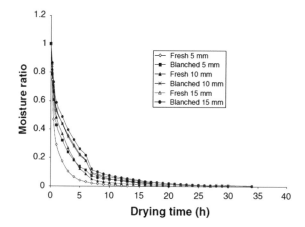

Fig. 6.7 Moisture ratio of sweet potato slices dried at 60 °C as affected by pretreatment and thickness (Falade and Solademi 2010)

Chiewchan et al. (2010) reported that blanching prior to drying of cabbage and spring onion affected the samples microstructure and the drying time. They concluded that blanching of cabbage in boiling water for 4 min resulted in maintenance of surface structure during drying but also led to higher volumetric shrinkage. For spring onion, they found that blanching was not able to deliver a proper surface appearance to the dried product due to surface shrinkage, which was accompanied by volumetric shrinkage. Drying rates were found to be higher for blanched vegetables when compared to unblanched ones, which was attributed to tissue softening during blanching. They used micrographs and the parameters roughness factor, shrinkage, moisture ratio and water activity to support their findings.

Differently from what would be expected, blanching before hot air drying of sweet potato slices increased drying time (Fig. 6.7) and reduced the values of effective moisture diffusivity (Falade and Solademi 2010). Specifically, blanching in water at 100 °C for 2 min resulted in effective moisture diffusivity between 6.36×10^{-11} and 1.78×10^{-9} m^2 s^{-1}, while unblanched samples presented values between 1.25×10^{-10} and 9.75×10^{-9} m^2 s^{-1} for this parameter. The authors attributed this behavior to starch gelatinization during blanching that impaired moisture movement during drying.

For the drying of Indian gooseberry or aonla, a previous blanching treatment was found to drastically reduce drying time (Gudapaty et al. 2010). Notwithstanding, the sensory quality, the vitamin C content and the rehydration capacity were higher for the unblanched product. The hypotheses of the authors for the decrease in vitamin C after blanching were leaching, diffusion, and thermal degradation. Another study on the drying of aonla shreds presented similar results, i.e., increase in drying rates when blanching was used (Gupta et al. 2014). The drying constants of various thin-layer drying models and the effective moisture diffusivity also increased with blanching.

The drying of wild pomegranate arils was accelerated by a blanching pretreatment in the study by Thakur et al. (2010). Steam blanching of arils for 30 s was claimed to disrupt their cell membrane thus facilitating water removal during drying. When blanching was followed by Sulfur fumigation at 0.3% during 60 min and drying at 60 °C in a cabinet drier, dried pomegranate arils of high sensory quality were obtained, i.e., good color, texture, aroma, taste, and overall acceptability. The authors attributed these results to the preserving effect of sulphuring on color and to the fast drying process achieved by means of blanching, thereby resulting in preservation of taste and aroma.

A report on eggplants also showed a decrease in hot air drying time with the use of blanching (Doymaz 2011). In that study, the eggplants were sliced to 1.5 cm and dried at 50, 60, 70, and 80 °C. Before drying, the slices were blanched in water at 70 °C for 3 min, while control samples were not blanched. It was clear for all temperatures tested that blanching reduced the drying time. This decrease reached 2 h for a drying temperature of 60 °C which is a significant value taking into consideration the usual times for hot air drying. These results impacted the values of effective moisture diffusivity which were higher for the blanched samples.

The drying of sweet basil by means of hot air and heat-pump drying yielded increased values of effective moisture diffusivity and of drying constant of the modified Henderson model after blanching the leaves in boiling water for 1 min (Phoungchandang and Kongpim 2012). In addition, the study showed that blanching helped in product color preservation, which was attributed to the inhibition of enzymatic browning. Furthermore, they found that blanching followed by heat-pump drying at 40 °C resulted in the best preservation of phenolics, the highest antioxidant capacity and the best rehydration ratio, which was related to lower drying temperatures and shorter drying times when blanching was used that ultimately resulted in better preservation of the sweet basil leaves cells.

The impact of blanching on the sun drying performance and on quality attributes of green pepper berries was studied by Gu et al. (2013), who found that blanching from 80 to 100 °C for 1–10 min led to quicker sun drying. They attributed this effect of rupture of the plant cells that accelerated the escape of moisture. Surprisingly, increase in blanching temperature and time resulted in more severe color changes, which suggests that other mechanisms besides enzymatic browning may affect the color of green pepper berries.

Doymaz (2014) studied the effect of blanching temperature and blanching time on drying parameters of broccoli. The author found that an increase in blanching temperature from 20 to 80 °C promoted an increase in drying rates and a decrease in drying time (Fig. 6.8), besides an improvement in rehydration capacity. In addition, he found that an increase in blanching time from 1 to 2 min caused an increase in drying rates and a decrease in drying time as well. Finally, he observed that higher values of effective moisture diffusivity were obtained for higher blanching temperatures (Fig. 6.9).

The drying of cocoyam slices at 50–70 °C in a convective drier was found to occur faster when preceded by blanching at 100 °C for 5 min, although blanching was inferior to the immersion in sodium meta-bisulphite for 5 min for this purpose (Afolabi et al. 2015).

The drying of apples was performed in two stages, viz. freeze-drying and hot air drying, being the first preceded by blanching at 100 °C for 1 min (Antal et al. 2015).

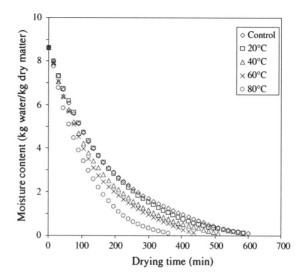

Fig. 6.8 Effect of blanching temperatures at 20, 40, 60 and 80 °C for 1 min on drying time (Doymaz 2014)

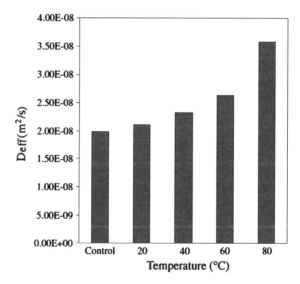

Fig. 6.9 Variation of effective diffusivity with different blanching temperature (Doymaz 2014)

The authors found that this combination of processes was more cost-effective when compared to freeze-drying and provided the product with better sensory quality when compared to hot air drying. In this case, the single use of blanching before freeze-drying reduced drying time in 21.74%. When these processes were combined with hot air drying, the decrease in drying time reached 60.87%.

The hot air drying of garlic as preceded by blanching in steam at 100 °C for 4 min was found to be quicker than the drying of unblanched samples (Fante and Noreña 2015). The authors confirmed this behavior by means of steeper drying curves and steeper water activity curves, along with higher values of the Henderson–Pabis and the Newton models constants. Additionally, the color of the blanched garlic was darker and yellower than that of the unblanched, which could indicate that blanching promoted the non-enzymatic browning reaction to some extent. Furthermore, the results showed that there was some leaching of sugars by blanching, even though it was performed with steam.

The drying of capelin fish was found to occur faster when blanching was used, but this pretreatment caused undesirable yellowness in the product, which made Cyprian et al. (2016) recommend the use of brining instead of blanching as a pretreatment for drying capelin.

Final Considerations

This chapter reported the use of blanching prior to drying focusing on, but not limiting to, the impact of this pretreatment on drying performance. In this sense, the blanching conditions should be optimized in order to reduce drying time and, at the same time, avoid severe losses in nutrient/bioactive content and sensory quality. It is important to test the use of blanching before drying for the food under study, since the observed effect is not always the same. However, blanching has commonly proved favorable for the performance of subsequent drying with acceptable impact on product quality.

References

Afolabi TJ, Tunde-Akintunde TY, Adeyanju JA (2015) Mathematical modeling of drying kinetics of untreated and pretreated cocoyam slices. J Food Sci Technol 52:2731–2740. Doi:10.1007/s13197-014-1365-z

Antal T, Kerekes B, Sikolya L, Tarek M (2015) Quality and drying characteristics of apple cubes subjected to combined drying (FD Pre-Drying and HAD Finish-Drying). J Food Process Preserv 39:994–1005. Doi:10.1111/jfpp.12313

Arévalo-Pinedo A, Murr FEX (2006) Kinetics of vacuum drying of pumpkin (*Cucurbita maxima*): modeling with shrinkage. J Food Eng 76:562–567. Doi:10.1016/j.jfoodeng.2005.06.003

Arévalo-Pinedo A, Xidieh Murr FE (2007) Influence of pre-treatments on the drying kinetics during vacuum drying of carrot and pumpkin. J Food Eng 80:152–156. Doi:10.1016/j.jfoodeng.2006.05.005

Bala BK, Hoque MA, Hossain MA, Uddin MB (2010) Drying characteristics of asparagus roots (*Asparagus racemosus Wild.*). Dry Technol 28:533–541. Doi:10.1080/07373931003618899

Blasco M (2006) Effect of blanching and air flow rate on turmeric drying. Food Sci Technol Int 12:315–323. Doi:10.1177/1082013206067352

Chiewchan N, Praphraiphetch C, Devahastin S (2010) Effect of pretreatment on surface topographical features of vegetables during drying. J Food Eng 101:41–48. Doi:10.1016/j.jfoodeng.2010.06.007

Ciurzyńska A, Lenart A, Greda KJ (2014) Effect of pre-treatment conditions on content and activity of water and colour of freeze-dried pumpkin. LWT Food Sci Technol. Doi:10.1016/j.lwt.2014.06.035

Cyprian O, Van Nguyen M, Sveinsdottir K et al (2016) Influence of lipid content and blanching on capelin (*Mallotus villosus*) drying rate and lipid oxidation under low temperature drying. J Food Process Eng 39:237–246. Doi:10.1111/jfpe.12215

Doymaz I (2014) Effect of blanching temperature and dipping time on drying time of broccoli. Food Sci Technol Int 20:149–157. Doi:10.1177/1082013213476075

Doymaz İ (2008) Drying of leek slices using heated air. J Food Process Eng 31:721–737. Doi:10.1111/j.1745-4530.2007.00185.x

Doymaz İ (2011) Drying of eggplant slices in thin layers at different air temperatures. J Food Process Preserv 35:280–289. Doi:10.1111/j.1745-4549.2009.00454.x

Falade KO, Solademi OJ (2010) Modelling of air drying of fresh and blanched sweet potato slices. Int J Food Sci Technol 45:278–288. Doi:10.1111/j.1365-2621.2009.02133.x

Fante L, Noreña CPZ (2015) Quality of hot air dried and freeze-dried of garlic (*Allium sativum L.*). J Food Sci Technol 52:211–220. Doi:10.1007/s13197-013-1025-8

Górnicki K, Kaleta A (2007) Drying curve modelling of blanched carrot cubes under natural convection condition. J Food Eng 82:160–170. Doi:10.1016/j.jfoodeng.2007.02.002

Gu F, Tan L, Wu H et al (2013) Analysis of the blackening of green pepper (*Piper nigrum Linnaeus*) berries. Food Chem 138:797–801. Doi:10.1016/j.foodchem.2012.11.033

Gudapaty P, Indavarapu S, Korwar GR et al (2010) Effect of open air drying, LPG based drier and pretreatments on the quality of Indian gooseberry (aonla). J Food Sci Technol 47:541–548. Doi:10.1007/s13197-010-0093-2

Gupta RK, Sharma A, Kumar P et al (2014) Effect of blanching on thin layer drying kinetics of aonla (*Emblica officinalis*) shreds. J Food Sci Technol 51:1294–1301. Doi:10.1007/s13197-012-0634-y

Hatamipour MS, Hadji Kazemi H, Nooralivand A, Nozarpoor A (2007) Drying characteristics of six varieties of sweet potatoes in different dryers. Food Bioprod Process 85:171–177. Doi:10.1205/fpb07032

Hossain MA, Woods JL, Bala BK (2007) Single-layer drying characteristics and colour kinetics of red chilli. Int J Food Sci Technol 42:1367–1375. Doi:10.1111/j.1365-2621.2006.01414.x

Kadam DM, Samuel DVK, Parsad R (2006) Optimisation of pre-treatments of solar dehydrated cauliflower. J Food Eng 77:659–664. Doi:10.1016/j.jfoodeng.2005.07.027

Kingcam R, Devahastin S, Chiewchan N (2008) Effect of starch retrogradation on texture of potato chips produced by low-pressure superheated steam drying. J Food Eng 89:72–79. Doi:10.1016/j.jfoodeng.2008.04.008

Kowalska H, Lenart A, Leszczyk D (2008) The effect of blanching and freezing on osmotic dehydration of pumpkin. J Food Eng 86:30–38. Doi:10.1016/j.jfoodeng.2007.09.006

Leeratanarak N, Devahastin S, Chiewchan N (2006) Drying kinetics and quality of potato chips undergoing different drying techniques. J Food Eng 77:635–643. Doi:10.1016/j.jfoodeng.2005.07.022

Molina Filho L, Frascareli EC, Mauro MA (2016) Effect of an Edible Pectin coating and blanching pretreatments on the air-drying kinetics of pumpkin (*Cucurbita moschata*). Food Bioprocess Technol 9:859–871. Doi:10.1007/s11947-016-1674-5

Mujumdar AS (ed) (2014) Handbook of industrial drying. CRC Press, Boca Raton

Paredes Escobar M, Gómez Galindo F, Wadsö L et al (2007) Effect of long-term storage and blanching pre-treatments on the osmotic dehydration kinetics of carrots (*Daucus carota* L. cv. Nerac). J Food Eng 81:313–317. Doi:10.1016/j.jfoodeng.2006.11.005

Phoungchandang S, Kongpim P (2012) Modeling using a new thin-layer drying model and drying characteristics of sweet basil (*Ocimum Baslicum Linn.*) using tray and heat pump-assisted dehumidified drying. J Food Process Eng 35:851–862. Doi:10.1111/j.1745-4530.2010.00633.x

Pimpaporn P, Devahastin S, Chiewchan N (2007) Effects of combined pretreatments on drying kinetics and quality of potato chips undergoing low-pressure superheated steam drying. J Food Eng c:318–329. Doi:10.1016/j.jfoodeng.2006.11.009

Scher CF, De Oliveira Rios A, Noreña CPZ (2009) Hot air drying of yacon (*Smallanthus sonchifolius*) and its effect on sugar concentrations. Int J Food Sci Technol 44:2169–2175. Doi:10.1111/j.1365-2621.2009.02056.x

Soria AC, Corzo-Martínez M, Montilla A et al (2010) Chemical and physicochemical quality parameters in carrots dehydrated by power ultrasound. J Agric Food Chem 58:7715–7722. Doi:10.1021/jf100762e

Thakur NS, Bhat MM, Rana N, Joshi VK (2010) Standardization of pre-treatments for the preparation of dried arils from wild pomegranate. J Food Sci Technol 47:620–625. Doi:10. 1007/s13197-010-0091-4

Tunde-Akintunde TY, Afolabi TJ (2009) Drying of chili pepper (*Capscium frutscens*). J Food Process Eng 33:649–660. Doi:10.1111/j.1745-4530.2008.00294.x

Vengaiah PC, Pandey JP (2007) Dehydration kinetics of sweet pepper (*Capsicum annum* L.). J Food Eng 81:282–286. Doi:10.1016/j.jfoodeng.2006.04.053

Walde SG, Velu V, Jyothirmayi T, Math RG (2006) Effects of pretreatments and drying methods on dehydration of mushroom. J Food Eng 74:108–115. Doi:10.1016/j.jfoodeng.2005.02.008

Workneh TS, Zinash A, Woldetsadik K (2014) Blanching, salting and sun drying of different pumpkin fruit slices. J Food Sci Technol 51:3114–3123. Doi:10.1007/s13197-012-0835-4

Yong CK, Islam MR, Mujumdar AS (2006) Mechanical means of enhancing drying rates: effect on drying kinetics and quality. Dry Technol 24:397–404. Doi:10.1080/07373930600616678

Chapter 7
Novel Blanching Techniques

Felipe Richter Reis

Abstract As stated in previous chapters, a conventional blanching treatment consists in immersing food pieces, usually vegetables, in hot water with the main goal of inactivating enzymes. Nevertheless, this treatment presents some drawbacks such as the high production of effluents and the leaching of nutrients. The use of steam for blanching foods might be considered as an innovation, since the afore-mentioned drawbacks are less pronounced than in hot water blanching. Nevertheless, the use of thermal treatments for fruits, for example, provides them with a cooked flavor and texture that is not appreciated by many consumers seeking for fresh-like fruit products. In this sense, this chapter brings studies on innovations on blanching techniques, from improvements on typical blanching processes to substitution of blanching by cutting edge techniques for enzymes inactivation. Technologies dealt with in the chapter include, for example: high-pressure pro-cessing, ultrasound, microwave blanching, Ohmic blanching, infrared blanching, and radiofrequency blanching. Novel blanching techniques will be presented sep-arated by type and chronologically in order for the reader to understand the advances in the each technology over the years.

Keywords High pressure · Ultrasound · Microwave blanching

In the beginning of the last decade, the impact of the use of recycled water on ascorbic acid retention during blanching of potato cylinders was studied by Arroqui et al. (2001). They found that using water containing 2.6 kg/m^3 of soluble solids, a concentration corresponding approximately to the one reached in recycled water of a pilot plant blancher after 2.5 h, increased the ascorbic acid retention during blanching at 80 °C from 10 to 60 min. Such result was confirmed by reduced values of apparent diffusivity of ascorbic acid when compared to blanching in distilled water. Their subsequent study (Arroqui et al. 2002) showed that an

F. Richter Reis (✉)
Food Technician Course, Instituto Federal do Paraná, Campus Jacarezinho, Jacarezinho, Paraná, Brazil
e-mail: felipe.reis@ifpr.edu.br

© Springer International Publishing AG 2017
F. Richter Reis (ed.), *New Perspectives on Food Blanching*,
DOI 10.1007/978-3-319-48665-9_7

increase in blanching time from 10 to 60 min and an increase in blanching temperature from 65 to 93 °C produce a decrease in ascorbic acid retention. Apparent diffusivity of ascorbic acid also increased with an increase in temperature. When using recycled water, though, the ascorbic acid retention was increased in up to 24.5% and the apparent diffusivity of ascorbic acid was reduced in up to 17%. These studies confirm the suitability of the use of recycled water as an innovation to reduce the losses of ascorbic acid during blanching of potatoes.

The use of high pressure as a substitute or as a complement of conventional blanching has been tested by several authors over the last years. According to Smith (2011), "high-pressure processing is a technique with the potential to preserve food without the use of the high temperatures normally employed in conventional thermal processing and therefore with considerable retention of quality attributes". Figure 7.1 shows the schematic of a batch high-pressure processing which is used for foods that cannot be pumped, also known as high hydrostatic pressure processing.

Al-Khuseibi et al. (2005) studied the impact of high hydrostatic pressure and hot water blanching on drying kinetics and quality parameters of hot air dried potato cubes. They found that potato treated by high pressure dried faster than hot water blanched potato, which could be attributed to increased permeability of cell structure obtained by means of high-pressure treatment. With concern to quality attributes of dried potatoes, their results show that high-pressure treated samples presented similar rehydrability, lower total color difference as compared to fresh samples, higher hardness, and higher density than blanched samples. They concluded that high hydrostatic pressure treatment instead of hot water blanching prior to hot air drying produces dried potatoes of better quality in a lower drying time.

The effect of high hydrostatic pressure on the enzymatic activity of frozen samples of enzyme extracts and pieces of potato, tomato, and carrot was assessed by Buggenhout et al. (2006). They found that the enzymes pectin methyl esterase (PME), polygalacturonase (PG), peroxidase (POD), and polyphenol oxidase (PPO) were not inactivated by the high-pressure treatment, while lipoxygenase (LOX) was pressure

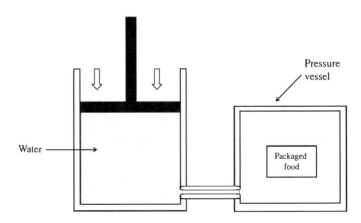

Fig. 7.1 Batch high-pressure processing (Smith 2011)

labile at pressures higher than 200 MPa. They concluded that, for the process conditions studied, a blanching pretreatment could not be substituted by a high-pressure treatment for inactivating most of the enzymes contained in the assessed products.

High hydrostatic pressure processing of swede (*Brassica napus*) was studied by Clariana et al. (2011). They found that the use of a 600 MPa pressure and temperature of 20 °C was as good for the quality of swede cylinders as blanching at 90 °C for 3 min. They expressed product quality in terms of texture, color, expressible moisture, i.e., moisture loss under compression, antioxidant activity, and concentration of bioactive compounds, viz. ascorbic acid, phenols, and glucosinolates. Their results suggest that high-pressure treatment at 600 MPa and 20 °C cause low texture degradation, low color changes, low moisture loss under compression, low losses of antioxidant capacity, and bioactive compounds, besides increase in the concentration of the glucosinolate progoitrin, which was attributed to hydrolysis of other glucosinolates induced by cell disruption during high-pressure processing. On the other hand, they observed that high-pressure processing at 400 MPa caused undesirable alterations in the characteristics of swede.

González–Cebrino et al. (2012) investigated the effect of high hydrostatic pressure in comparison to pasteurization, both preceded or not by blanching, on the quality of plum puree after 20 d of refrigerated storage. They found that microbiological counts were lower for the thermally treated samples, as expected. Total soluble solids, pH, and titratable acidity were roughly unaffected in all samples. On the other hand, PPO activity was lower in thermally treated samples, which resulted in a more preserved color. As for the carotenoids, both blanching and high pressure seem to improve their extractability. The same behavior was observed for phenolic compounds, thereby positively impacting on the samples antioxidant activity. They concluded that a blanching step is necessary before high-pressure processing of plum puree in order to reduce PPO activity and thereby preserve color.

The high-pressure processing of white cabbage was studied by Alvarez–Jubete et al. (2014), who found that it was superior to blanching for bioactive compounds preservation and antioxidant activity. However, blanching was better for color and texture preservation. Blanching was performed at 90–95 °C for 3 min, while high pressure was carried out at 200–600 MPa for 5 min at a temperature of 20 or 40 °C. The best high-pressure treatment conditions were a pressure of 600 MPa and temperature of 20 or 40 °C. The negative impact of blanching on bioactive compounds and antioxidant activity may be attributed to leaching and thermal destruction. The negative impact of high pressure on color and texture may be attributed to changes in structural characteristics.

Blanching is a usual process during processing of pumpkin. In order to find whether high-pressure processing could be better than blanching and pasteurization for the quality of pumpkin puree, Contador et al. (2014) studied the effect of these processes on pumpkin puree physicochemical parameters, color, phenolic compound, and carotenoids concentration. They found that physiochemical parameters were not severely affected by process conditions, while thermal treatments reduced the color intensity of the puree. The authors hypothesized that the increased bioavailability of pigments in thermally treated puree may have resulted in increased exposure of them to degradation reactions, which justifies the results of color measurements. Furthermore, the study showed that bioactive compounds were more preserved by using high-pressure treatments. The negative impact of pasteurization on bioactive compounds may be attributed to the degradation effect of heat on these compounds. Even though thermal treatments were usually inferior to high pressure in terms of product quality, the authors recommended that the use of blanching combined with acidification would be proper for providing high-pressure treated pumpkin puree with color stability during storage thereby resulting in an extended shelf-life.

García–Parra et al. (2014) investigated the effect of blanching and ascorbic acid pretreatments as combined with pasteurization and high-pressure processing on the quality of nectarine puree. They found that blanching was more suitable when combined with high-pressure processing, while ascorbic acid addition was better when combined with pasteurization. Both of these strategies were able to maintain the microbial counts below the detection limit. As for PPO activity, they observed that either a thermal pretreatment (blanching) or a thermal treatment (pasteurization) were necessary for reducing PPO activity to acceptable levels after processing (Fig. 7.2) and after 30 d of refrigerated storage (Fig. 7.3). These results reflected on color measurements. They also found that carotenoids and phenols were better

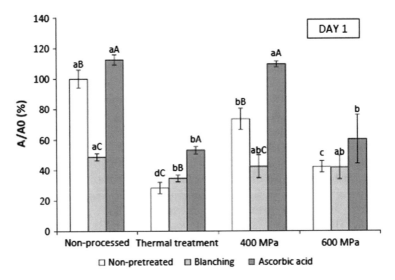

Fig. 7.2 Changes in the polyphenol oxidase activity ($A/A_0 \times 100$) after processing [thermal and high-pressure processing (HPP)] of a nectarine puree manufactured at different conditions (nonpretreated, blanching, ascorbic acid addition). *a, b, c, d* Different *lower-case letters* in purees with the same pretreatment (columns with the same color) indicate significant differences due to pasteurization process (Tukey's test, $P < 0.05$). *A, B, C* Different capital letters in the same group of column indicate significant statistical differences due to initial manufacture conditions (Tukey's test, $P < 0.05$) (García-Parra et al. 2014)

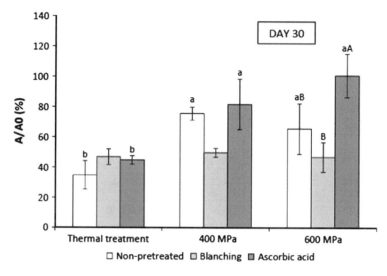

Fig. 7.3 Changes in the polyphenol oxidase activity ($A/A_0 \times 100$) after 30 days of refrigerated storage of nectarine puree manufactured at different conditions (nonpretreated, blanching, ascorbic acid addition). *a, b* Different *lower-case letters* in purees with the same pretreatment (columns with the same color) indicate significant statistical differences due to pasteurization process (Tukey's test, $P < 0.05$). *A, B* Different capital letters in the same group of column indicate significant statistical differences due to initial manufacture conditions (Tukey's test, $P < 0.05$) (García-Parra et al. 2014)

preserved when using blanching—high pressure or ascorbic acid—pasteurization. In conclusion, a thermal processing step showed necessary during the processing of nectarine puree for providing the product with shelf-stability. Ascorbic acid in combination with high pressure was not effective in this sense.

Ultrasound is another technique that has been extensively studied over the last ten years with regard to its impact on food quality. It uses frequencies between 20 and 100 kHz and power levels between 10 and 1000 W cm^{-2} to generate vibrations to produce cavitation, a phenomenon in which bubbles are formed and then collapse generating localized pressures of the order of 100 MPa, temperatures as high as 5000 K, and forces enough to damage cell walls in foods of high moisture content. Besides reducing the heat required for thermal treatments, the use of ultrasound improves the heat transfer (Smith 2011).

The influence of high-pressure processing and ultrasound treatment, alone or in combination, on quality characteristics of carrot juice, was studied by Jabbar et al. (2014b). Regarding enzyme inactivation, the POD, PPO, PME, and LOX activities were more drastically reduced by means of hot water blanching. However, significant reductions on enzymatic activities were also observed for samples treated by a combination of high-pressure processing and ultrasound, which was attributed to a synergistic effect between these methods thereby affecting the functional groups of the enzymes (Table 7.1). In addition, they observed that phenols, flavonoids, tannins, ascorbic acid, and carrot carotenoids were best preserved by

Table 7.1 Effects of US and HHP on POD, PPO, PME, and LOX residual activity percentage in carrot juice ($n = 3$) (Jabbar et al. 2014b)

Treatments	POD (residual activity %)	PPO (residual activity %)	POD (residual activity %)	PME (residual activity %)
Control	100.00 ± 0.00a	100.00 ± 0.00a	100.00 ± 0.00a	100.00 ± 0.00a
WB	17.75 ± 0.26h	13.89 ± 0.15h	14.78 ± 0.28g	12.74 ± 0.36h
US	99.98 ± 0.11a	99.97 ± 0.22a	99.96 ± 0.22a	99.98 ± 0.21a
HHP250	78.13 ± 0.24b	69.99 ± 0.24b	74.37 ± 0.15b	72.99 ± 0.19b
HHP350	65.72 ± 0.26c	58.11 ± 0.16c	63.36 ± 0.27c	61.56 ± 0.24c
HHP450	49.31 ± 0.25f	45.41 ± 0.25f	47.81 ± 0.14e	46.13 ± 0.17f
US-HHP250	63.26 ± 0.17d	53.18 ± 0.17d	58.45 ± 0.16d	55.17 ± 0.35d
US-HHP350	52.81 ± 0.14e	45.96 ± 0.13e	47.85 ± 0.13e	48.79 ± 0.23e
US-HHP450	33.31 ± 0.16g	29.10 ± 0.15g	32.05 ± 0.25f	31.09 ± 0.26g

Values with different letters in the same column (a–h) are significantly different ($P < 0.05$) from each other. *WB* water blanched, *US* ultrasound, *HHP250* high hydrostatic pressure at 250 MPa, *HHP350* high hydrostatic pressure at 350 MPa, *HHP450* high hydrostatic pressure at 450 MPa, *US-HHP250* ultrasound + high hydrostatic pressure at 250 MPa, *US-HHP350* ultrasound + high hydrostatic pressure at 350 MPa, *US-HHP450* ultrasound + high hydrostatic pressure at 450 MPa

high-pressure processing combined with ultrasound. They hypothesize that the cavitation effect of ultrasound together with the harm caused to the cells by high pressure helped to release the bioactive compounds into the liquid. High pressures were found to be necessary to prevent the enzymatic browning of the juice and thereby preserve its color. The authors concluded that high-pressure processing at 450 MPa combined with ultrasound is a feasible way to process carrot juice.

Cruz et al. (2006) found that thermosonication, i.e., a combination of ultrasound and heat treatments, promoted inactivation of peroxidase in watercress more effectively than conventional hot water blanching, suggesting that it can be an alternative in which lower thermal processing times are used, ultimately resulting in a more preserved product. The use of conventional blanching as compared to thermosonication before freezing of watercress was studied by Cruz et al. (2007). They found that the use of thermosonication led to greener color in the samples when compared to raw and blanched samples. In conclusion, they observed that thermosonication prior to freezing is a feasible way to increase the shelf-life of watercress without pigment degradation. The same group of researchers (Cruz et al. 2008) observed that vitamin C losses after thermosonication were much lower (6%) than after thermal blanching alone (71%), which was attributed to a 14-fold reduction in process time when comparing the methods.

A study on miscellaneous foods was conducted by Alexandre et al. (2011), who found that thermosonication between 50–65 °C, at a frequency of 35 kHz, and 120 W of power was as effective as blanching, for similar process temperatures, for reducing the microbial counts in red bell pepper, watercress, and strawberry. Reductions of up to 8 logs were achieved in the study. At the same time, they verified that thermosonication preserved the quality of the product better than blanching. The bactericidal effect of thermosonication was attributed to heat and to cavitation, the latter promoting disruption of the cellular structure. The same study also showed that ultraviolet radiation was not an efficient method for reducing the microbial counts in the samples, although it preserved well the color, the texture, and the anthocyanins of the assessed foods.

Ultrasound pretreatment at 25 °C, frequency of 20 kHz, amplitudes between 24.4 and 61.0 μm, and process times between 3 and 10 min with pulses of 5 s was found effective for producing dried carrots of good color and preserved bioactive content (Rawson et al. 2011). Ultrasound was superior to hot water blanching at 80 °C for 3 min with regard to lightness (L*), retention of carotenoids and selected polyacetylenes. The lower L* value in heat treated carrots may be attributed to nonenzymatic browning, while the color losses during hot air drying may be due to carotenoid oxidation. Additionally, blanching may have promoted thermal degradation of the bioactive compounds.

The use of ultrasound was compared to the use of blanching with regard to its impact on the texture and microstructure of retorted carrot disks (Day et al. 2012). The authors found that both pretreatments provided protection to the middle lamella by avoiding the breakdown of pectins and thereby providing the samples with higher values of resistance to compression and tensile forces. Figure 7.4 shows the microstructure of carrot disks after different pretreatments and retorting, where it can be seen that untreated samples present a middle lamella more affected by retorting, i.e., more separated and deformed cells. The use of CaCl₂ in the pretreatment water promoted a strengthening effect on carrot texture, which can be attributed to the formation of calcium pectate. Furthermore, blanching at low temperatures like the one used in the study (60 °C) is known to activate PME in the tissue thus promoting a strengthening effect in vegetable tissues, as discussed in Chap. 2.

Another work on ultrasound pretreatment showed that thermosonication was superior to thermal treatment for the inactivation of PPO in mushrooms

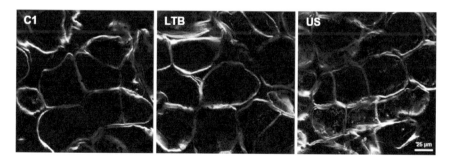

Fig. 7.4 Micrographs of the carrot cylinders after retorting showing differences in cell structure depending on the specific pretreatment: *C1* the control (no pretreatment), *LTB* low temperature long time blanching (60 °C, 40 min), and *US* low temperature with the application of ultrasound (60 °C, 10 min) (Day et al. 2012)

(Cheng et al. 2013). The authors came to this conclusion on the strength of higher values of inactivation constant (K, min^{-1}), lower values of decimal reduction time (D, min), and lower values of half-life ($t_{1/2}$, min^{-1}) obtained for the thermosonication pretreatment. Thermosonication was performed at 55–75 °C, 25 kHz frequency, and 600 W of power with pulses of 5 s followed by 5 s off. Thermal treatment of the enzyme extracts were performed at the same temperatures used during thermosonication for times up to 30 min. An interesting conclusion of the study was that the inactivation effect of thermosonication was more pronounced at lower temperatures, which was attributed to the lower vapor pressure at low temperatures thus allowing a more violent cavitation.

With regard to the quality of processed carrots, Gamboa-Santos et al. (2013) found that High Temperature Short Time (HTST) blanching was better than ultrasound for the preservation of carrot vitamin C. More specifically, their results show that blanching of carrot slices in boiling water for 1 min or in steam for 2 min resulted in vitamin C retentions of 85% and 81.2%, respectively, which were the highest among all pretreatments. Ultrasound pretreatment combined with heating at 60–70 °C resulted in severe losses of vitamin C, which were attributed to the formation of small channels due to cavitation which ultimately facilitated the transport of nutrients outside the samples.

Ultrasound pretreatment combined with blanching (thermosonication) was considered better than blanching alone for inactivating POD while preserving chlorophyll in *Artemisia argyi* leaves (Xin et al. 2013). Thermosonication resulted in high values of reaction rate constant for the inactivation of POD, being these values higher for higher ultrasound power levels and for higher water temperatures. They concluded that an ultrasonic intensity of 11.94 W/cm^2 at 85 °C for 60 s was the best strategy for obtaining high POD inactivation (92.7%) while preserving chlorophyll (96.7% of retention).

Jabbar et al. (2014a) evaluated the effect of blanching and ultrasound on quality features of carrot juice. The equipment used is shown in Fig. 7.5. Such equipment was operated in pulsed mode (5 s on, 5 s off) at a frequency of 20 kHz, an amplitude level of 70%, and an ultrasonic intensity of 48 W/cm^2. Blanching of the carrots was performed either in pure water or in acidified water at 100 °C for 4 min. They found that the blanching of the carrots followed by ultrasound treatment of carrot juice caused a positive impact on carrot juice quality, namely high level of chlorogenic acid, total carotenoids, lycopene and lutein, low total plate counts, and low yeast and mold counts, along with low losses of sugars and minerals. The effects of heat and cavitation may have increased the availability of the bioactive compounds and promoted killing of microorganisms. Another study conducted by the same research group (Jabbar et al. 2014c) showed that combining blanching and sonication is also good for the inactivation of POD and for the preservation of carrot juice original color and selected bioactive compounds, along with high values of antioxidant activity.

Recently, Amaral et al. (2016) found that ultrasound pretreatment was better than blanching for the color and for the texture of vacuum-packaged potato strips before and after frying, viz., lower color changes, and lower firmness loss.

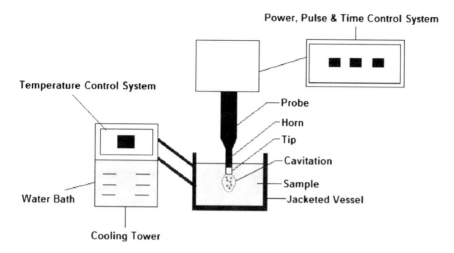

Fig. 7.5 Schematic diagram of probe type sonication exposure system (Jabbar et al. 2014a)

They concluded that the use of ultrasound at 40 kHz, 200 W for 3 min is a promising technology for substituting blanching in the potato industry.

Microwave blanching is another extensively studied technology. The use of microwaves for heating foods presents the advantage of fast heating. This, indeed, can also be the drawback of this technology, since fast heating can cause the food to easily heat-up to temperatures that cause burning. Nevertheless, the advantages of microwave heating seem to surpass the drawbacks since this technology is increasingly present in homes and food industries. According to Ranjan et al. (2016), microwave blanching heats materials that have a dielectric medium by means of the alignment of this medium towards the oscillating electromagnetic field. They affirm that molecules that have an electric dipole moment experience molecular rotation that align them in an electromagnetic field, which increases the kinetic energy and thereby, product temperature.

Severini et al. (2005) studied the blanching of potato cubes prior to drying, finding that microwave blanching was better than conventional blanching as a pretreatment to drying. Microwave blanching was performed by immersing the cubes in distilled water or in 3% NaCl containing water and placing these systems in a microwave oven at 850 W for 5 min. The use of microwave blanching followed by belt drying resulted in a product with better color and rehydration properties which was obtained in a lower drying time.

Blanching of peanuts by using microwaves was studied by Schirack et al. (2006a). In this case, blanching was not performed in water, but inside a 915 MHz microwave tunnel unit with a belt conveyor equipped with a fan. Samples were blanched from 4 to 11 min at a power of 5 kW. They found that the higher the blanching time, the higher the energy absorbed by the peanuts. The use of a fan resulted in lower product temperatures, even though the temperatures reached were above those attained during conventional oven blanching of peanuts. In addition, peanuts with higher initial moisture content heated less than those with lower initial moisture content. Higher blanching times also resulted in higher blanchability, i.e., lower percentage of peanuts with skin attached. Furthermore, blanchability was higher for samples with lower initial moisture content. They concluded that microwave blanching is a suitable technique for blanching peanuts, especially taking into account the reduced process time required. Nevertheless, another study by the same authors suggests that microwave blanching can generate off-flavors in peanuts due to the high temperature attained in the product during the process which favors thermal degradation and Maillard reactions (Schirack et al. 2006b).

A study on the quality of Brussels sprouts tested two pretreatment conditions: one of them consisted in immersion in water at 50 °C for 5 min and the other comprised heating in a microwave oven at 700 W for 5 min (Viña et al. 2007). Both were followed by blanching in boiling water, being the former for 3 min and the latter for 2 min. Blanching in boiling water alone during 1, 3, or 4 min was also tested. They observed that microwave blanching promoted the lowest lightness loss among all pretreatments tested. Furthermore, microwave blanching yielded acceptable texture changes, chlorophyll, and flavonoids losses, besides increasing the ascorbic acid extractability and the radical scavenging activity of the samples. The authors concluded that a microwave step during blanching of Brussel sprouts was a good option for preserving their quality prior to other processes, such as freezing.

Dorantes-Alvarez et al. (2011) blanched pepper paste using a microwave oven, finding that the time to reach PPO inactivation temperature (80 °C) was 15 s. The process time for inactivation when such temperature was attained in the surface of the paste was 5 s. Inactivation was achieved by using an energy density of 0.38 kJ/g. They observed that the microwave blanched product antioxidant activity increased, while phenolic content decreased. This behavior was attributed to the generation of phenolic derivatives of higher antioxidant capacity.

Latorre et al. (2012) studied the effect of microwave blanching on quality features of red beet, finding that direct microwave blanching (without immersion of the product in water) resulted in high shrinkage and weight loss. On the other hand, they concluded that microwave blanching of the beets immersed in water is a feasible way to inactivate POD and PPO while preserving color and texture. Palma-Orozco et al. (2012) found that a microwave blanching treatment at 900 W during 165 s (power density of 0.51 kJ/g) was enough to inactivate PPO in *Pouteria sapota* fruit pulp. Bernaś and Jaworska (2014) compared the use of conventional blanching in water and in heated sodium metabisulphite/citric acid solution with microwave blanching alone (1000 W for 5 min) and with a combination of hot water (96–98 °C for 1 min) and microwave blanching (3 min). The effect of these pretreatments on the chemical composition, color, antioxidant, and enzymatic activity of mushrooms was assessed. They concluded that microwave blanching treatment alone was favorable to the preservation of dry matter, ash, vitamin B_1, and vitamin B_2 content along with PPO and POD inactivation. On the other hand, thermal–chemical blanching was found better for preserving the color of the mushrooms. Binsi et al. (2014) found that direct microwave blanching of sutchi catfish (*Pangasianodon hypophthalmus*) filets increased the shelf-life of the chilled product from 12 to 21 days while promoting only small changes on the content of fatty acids and minerals.

Microwave blanching of carrots was studied by Sezer and Demirdöven (2015), who found that the optimum conditions for this pretreatment were the immersion of 50 g samples in 75 mL of water during 300 s at 360 W of microwave power. The product obtained under these blanching conditions presented preserved pectin, carotenoids, dry matter, texture, and color, while conventionally blanched carrots (300 s/94 °C) presented lower pectin, carotenoid, and dry matter content.

Severini et al. (2016) compared the use of direct microwave, hot water, and steam blanching methods for the inactivation of POD in broccoli. In addition, color, ascorbic acid, and phenolics were used to express blanched products quality. The authors found that microwave and steam blanching were better than hot water blanching for inactivating POD in the tissue (Fig. 7.6). Their results also show that microwave blanched samples were yellower than steam and hot water blanched ones. Furthermore, they found that ascorbic acid was better preserved with the use of microwave blanching, while phenolics were roughly similarly preserved when using different blanching techniques. In conclusion, the authors found that microwave blanching may be a suitable, environmental friendly technology for preserving broccoli.

Infrared blanching technology has been tested for inactivating enzymes in foods as well. According to a review provided by Krishnamurthy et al. (2008), "infrared

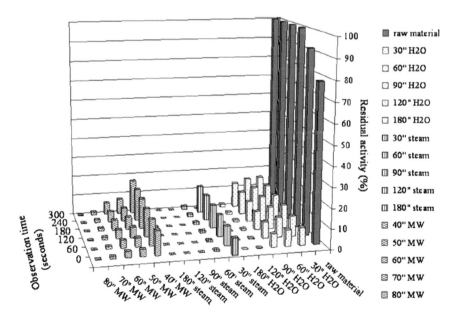

Fig. 7.6 Peroxidase inactivation as a function of heating times and type of treatment (Severini et al. 2016)

heating is gaining popularity because of its higher thermal efficiency and fast heating rate/response time in comparison to conventional heating." During infrared heating, heat is transferred to the food by radiation in the form of electromagnetic waves whose wavelength range from 0.78 to 1000 μm (Krishnamurthy et al. 2008).

Bingol et al. (2012) studied the influence of infrared dry-blanching, as performed at 11,080 W/m^2 for 30–180 s, on oil uptake of French fries during frying. They found that this type of blanching was able to reduce oil absorption in up to 37.5% during frying at 146 °C. They attributed this result to starch gelatinization, formation of a crust, and formation of a whitish elastic skin which prevented the penetration of oil into the samples. Additionally, the same study showed that infrared blanching inactivated POD and provided the French fries with superior sensory quality as compared to unblanched ones. Guiamba et al. (2015) evaluated the impact of infrared blanching on PPO and ascorbic acid oxidase (AAO) activities, and on the retention of bioactive compounds in dried mango. They selected infrared power outputs as to obtain a similar temperature history to water blanching at 65 °C for 10 min and 90 °C for 2 min. They found that infrared blanching led to inactivation or acceptable decrease in the activity of both enzymes, along with better preservation of ascorbic acid and acceptable retention of All-*trans*-β-carotene. In addition, infrared blanching accelerated the drying process.

Ohmic heating technology can be used for blanching foods with advantages like rapid and uniform heating. According to Smith (2011), "it is a method of heating food by passing an alternating electric current directly through it." A review by Knirsch et al. (2010) states that Ohmic heating is an internal thermal energy generation technology that does not depend on heat transfer either through a solid–liquid interface or inside a solid in a two-phase system.

The study by Icier et al. (2006) showed that Ohmic blanching was able to inactivate POD while preserving color of pea puree. They used different voltage gradients (20–50 V/cm) and frequency of 50 Hz to heat the puree from 30 to 100 °C. Conventional blanching in water at 100 °C was performed in parallel to Ohmic blanching. In conclusion, the authors pointed out that Ohmic blanching could be an alternative to hot water blanching for inactivating enzymes in pea puree, an effect that can be attributed to the denaturing effect of heat on one food enzymes. Lemmens et al. (2009) compared the effect of conventional, microwave, and Ohmic blanching on the quality of carrot pieces, finding that these different types of heating did not affect differently the product enzymatic activity, carotenoid content, and degree of methoxylation. Guida et al. (2013) found that Ohmic heating provided faster inactivation of PPO and POD in artichokes when compared to conventional blanching. The apparatus used by the authors is shown in Fig. 7.7. The samples were Ohmic blanched in a water/vinegar solution at a gradient voltage of 24 V/cm. The study showed that Ohmic blanching yields samples with higher lightness, higher color saturation, and lower total color difference from unblanched samples, indicating that this method preserves the bright green color of raw artichokes. They also observed that Ohmic blanching preserved better the artichoke proteins and polyphenols, besides causing desirable uniform softening of the product.

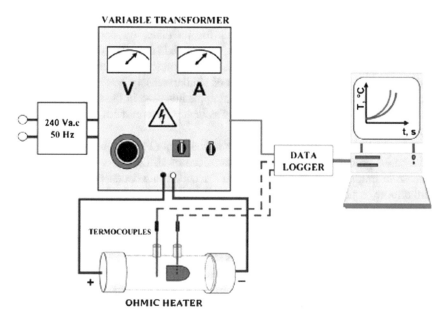

Fig. 7.7 Experimental set up of ohmic heating system (Guida et al 2013)

The Ohmic blanching of mussels (9.15 V/cm, 60 Hz) was compared to conventional blanching (50–90 °C) in terms of their effect on heavy metals concentration, texture, and microbial load of the product (Bastías et al. 2015). The authors found that Ohmic heating reduced the concentration of lead and cadmium, the cutting strength and the mesophilic aerobic and enterobacteriaceae counts, such as did hot water blanching. Ohmic blanching at 90 °C was especially good for reducing Cd concentration, yet it promoted severe softening of mussels. The effect of Ohmic heating on microorganisms may be attributed to heat and also to electroporation, i.e., the formation of pores in cell membranes due to the presence of an electric field (Knirsch et al. 2010).

Other attempts of innovation on blanching technology have been made over the last years. For example, the use of whirling bed blanching, i.e., "an improved version of fluidized bed blanching using a mixture of steam and hot air as the medium" (Mukherjee and Chattopadhyay 2007) was studied. The authors found that whirling bed blanching of potatoes at 85 °C promoted faster POD inactivation while retaining a higher level of reducing sugars and ascorbic acid when compared to hot water and steam blanching.

The use of high-pressure carbon dioxide for inactivating PPO and POD in red beet was tested by Liu et al. (2008), who found that this technology was able to inactivate the selected enzymes in red beet at significant higher inactivation rates as compared to thermal treatment. In addition, specific conditions of high-pressure carbon dioxide treatment (37.5 MPa, 55 °C, 60 min) were more favorable for the retention of red pigments than hot water blanching.

Radiofrequency blanching was assessed for the inactivation of PPO and LOX in model systems—enzyme solutions—and for its effect on apple puree quality (Manzocco et al. 2008). The authors concluded that the thermal effect of radiofrequency inactivated both enzymes. Furthermore, radiofrequency blanching of the whole apples yielded apple puree of similar sensory quality to the conventionally blanched puree. The experimental apparatus used in the study consisted of a pilot-scale 3.5 kW, 27.12 MHz radiofrequency system with plate applicators operating at 4.0–6.9 kV.

The exposure to UV-C light in order to inactivate PPO in model systems and apple-based products was studied by Manzocco et al. (2009), who found that an increase in irradiance from 0 to 13.8 Wm^{-2} resulted in a decrease in PPO activity. Furthermore, exposure to visible light at 12.7 Wm^{-2} reduced PPO activity in about 70% after 16 h. The study showed that the use of light can be an alternative for blanching products that undergo undesirable enzymatic reactions but "cannot" be blanched such as fresh-cut fruit and vegetables. The combined use of UV-C light and heat for inactivating POD in zucchini was studied by Neves et al. (2012), who found that UV-C light enhance the inactivating effect of heat above 85 °C, thus allowing the use of shorter thermal blanching treatments. Stamatopoulos et al. (2012) observed no effect of UV-C light treatment on the content of oleuropein, a phenolic of olive leaves, while steam blanching increased its content from 25 to 35 times.

The use of superheated steam containing microdroplets of hot water for blanching potatoes was assessed by Sotome et al. (2009), who observed that such technology was suitable for inactivating POD and PPO while preserving color, texture, and weight of the samples. They suggest that superheated steam combined with hot water spray be used for other fruit and vegetables in an attempt to substitute conventional hot water blanching with advantages.

High humidity air impingement blanching technique was proposed for treating apple quarters by Bai et al. (2013a). According to the authors, "during the high humidity air impingement blanching process, the hot air with high humidity impinges on the product surface at high velocity, which can make the thermal boundary layers become thinner and increase the rate of heat transfer." They concluded that the proposed method as carried out at 90–120 °C was able to inactivate PPO while retaining vitamin C. Another work by the same authors assessed the high humidity air impingement blanching for accelerating the drying of seedless grapes, finding that this innovative blanching method was effective for reducing drying times while preserving product color (Bai et al. 2013b). They affirmed that optimized blanching conditions, viz., blanching at 110 °C for 90 s at 15 m/s air speed followed by drying at 60 °C gives origin to high quality dried seedless grapes.

Pressure blanching, i.e., blanching inside an autoclave was effective for treating peanuts in order to produce high quality peanut milk (Jain et al. 2013). Blanching at 121 °C at 15 psi for 3 min was found to be the best pretreatment in terms of peanut milk sensory and physicochemical quality. Additionally, the new method reduced the time of the mandatory peanut soaking treatment from 16 to 6 h, representing an advance from the economical viewpoint.

Final Considerations

As demonstrated in this chapter, new blanching methods have been developed over the years in an attempt to improve product quality and process efficiency. Several new methods continue under investigation, including the ones presented here, that are the most common alternatives to conventional hot water and steam blanching. The choice of the blanching method will depend on the capital available for investment and on specific food product characteristics. Regardless of the type of innovation, advances in blanching technology are relevant to the food industry and studies in this field are always welcome.

References

Alexandre EMC, Santos-Pedro DM, Brandão TRS, Silva CLM (2011) Study on thermosonication and ultraviolet radiation processes as an alternative to blanching for some fruits and vegetables. Food Bioprocess Technol 4:1012–1019. Doi:10.1007/s11947-011-0540-8

Al-Khuseibi MK, Sablani SS, Perera CO (2005) Comparison of water blanching and high hydrostatic pressure effects on drying kinetics and quality of potato. Dry Technol 23:2449–2461. Doi:10.1080/07373930500340734

Alvarez-Jubete L, Valverde J, Patras A et al (2014) Assessing the impact of high-pressure processing on selected physical and biochemical attributes of white cabbage (*Brassica oleracea* L. var. capitata alba). Food Bioprocess Technol 7:682–692. Doi:10.1007/s11947-013-1060-5

Amaral RDAA, Benedetti BC, Pujolà M et al (2016) A first approach of using ultrasound as an alternative for blanching in vacuum-packaged potato strips. Food Bioprocess Technol. Doi:10.1007/s11947-016-1758-2

Arroqui C, Rumsey T, Lopez a, Virseda P (2001) Effect of different soluble solids in the water on the ascorbic acid losses during water blanching of potato tissue. J Food Eng 47:123–126. Doi:10.1016/S0260-8774(00)00107-2

Arroqui C, Rumsey T, Lopez a, Virseda P (2002) Losses by diffusion of ascorbic acid during recycled water blanching of potato tissue. J Food Eng 52:25–30. Doi:10.1016/S0260-8774(01)00081-4

Bai JW, Gao ZJ, Xiao HW et al (2013a) Polyphenol oxidase inactivation and vitamin C degradation kinetics of Fuji apple quarters by high humidity air impingement blanching. Int J Food Sci Technol 48:1135–1141. Doi:10.1111/j.1365-2621.2012.03193.x

Bai JW, Sun DW, Xiao HW et al (2013b) Novel high-humidity hot air impingement blanching (HHAIB) pretreatment enhances drying kinetics and color attributes of seedless grapes. Innov Food Sci Emerg Technol 20:230–237. Doi:10.1016/j.ifset.2013.08.011

Bastías JM, Moreno J, Pia C et al (2015) Effect of ohmic heating on texture, microbial load, and cadmium and lead content of Chilean blue mussel (*Mytilus chilensis*). Innov Food Sci Emerg Technol 30:98–102. Doi:10.1016/j.ifset.2015.05.011

Bernaś E, Jaworska G (2014) Effect of microwave blanching on the quality of frozen Agaricus bisporus. Food Sci Technol Int. Doi:10.1177/1082013214529956

Bingol G, Zhang A, Pan Z, McHugh TH (2012) Producing lower-calorie deep fat fried French fries using infrared dry-blanching as pretreatment. Food Chem 132:686–692. Doi:10.1016/j.foodchem.2011.10.055

Binsi PK, Ninan G, Zynudheen AA et al (2014) Compositional and chill storage characteristics of microwave-blanched sutchi catfish (*Pangasianodon hypophthalmus*) fillets. Int J Food Sci Technol 49:364–372. Doi:10.1111/ijfs.12308

Buggenhout S, Messagie I, der Plancken I, Hendrickx M (2006) Influence of high-pressure–low-temperature treatments on fruit and vegetable quality related enzymes. Eur Food Res Technol 223:475–485. Doi:10.1007/s00217-005-0227-3

Cheng X-f, Zhang M, Adhikari B (2013) The inactivation kinetics of polyphenol oxidase in mushroom (*Agaricus bisporus*) during thermal and thermosonic treatments. Ultrason Sonochem 20:674–679. Doi:10.1016/j.ultsonch.2012.09.012

Clariana M, Valverde J, Wijngaard H et al (2011) High pressure processing of swede (*Brassica napus*): impact on quality properties. Innov Food Sci Emerg Technol 12:85–92. Doi:10.1016/j.ifset.2011.01.011

Contador R, González-Cebrino F, García-Parra J et al (2014) Effect of hydrostatic high pressure and thermal treatments on two types of pumpkin purée and changes during refrigerated storage. J Food Process Preserv 38:704–712. Doi:10.1111/jfpp.12021

Cruz RMS, Vieira MC, Silva CLM (2006) Effect of heat and thermosonication treatments on peroxidase inactivation kinetics in watercress (*Nasturtium officinale*). J Food Eng 72:8–15. Doi:10.1016/j.jfoodeng.2004.11.007

Cruz RMS, Vieira MC, Silva CLM (2007) Modelling kinetics of watercress (*Nasturtium officinale*) colour changes due to heat and thermosonication treatments. Innov Food Sci Emerg Technol 8:244–252. Doi:10.1016/j.ifset.2007.01.003

Cruz RMS, Vieira MC, Silva CLM (2008) Effect of heat and thermosonication treatments on watercress (*Nasturtium officinale*) vitamin C degradation kinetics. Innov Food Sci Emerg Technol 9:483–488. Doi:10.1016/j.ifset.2007.10.005

Day L, Xu M, Øiseth SK, Mawson R (2012) Improved mechanical properties of retorted carrots by ultrasonic pre-treatments. Ultrason Sonochem 19:427–434. Doi:10.1016/j.ultsonch.2011.10.019

Dorantes-Alvarez L, Jaramillo-Flores E, González K et al (2011) Blanching peppers using microwaves. Procedia Food Sci 1:178–183. Doi:10.1016/j.profoo.2011.09.028

Gamboa-Santos J, Cristina Soria a, Pérez-Mateos M et al (2013) Vitamin C content and sensorial properties of dehydrated carrots blanched conventionally or by ultrasound. Food Chem 136:782–788. Doi:10.1016/j.foodchem.2012.07.122

García-Parra J, Contador R, Delgado-Adámez J et al (2014) The applied pretreatment (blanching, ascorbic acid) at the manufacture process affects the quality of nectarine purée processed by hydrostatic high pressure. Int J Food Sci Technol 49:1203–1214. Doi:10.1111/ijfs.12418

González-Cebrino F, García-Parra J, Contador R et al (2012) Effect of high-pressure processing and thermal treatment on quality attributes and nutritional compounds of "Songold" plum purée. J Food Sci 77:C866–C873. Doi:10.1111/j.1750-3841.2012.02799.x

Guiamba IRF, Svanberg U, Ahrné L (2015) Effect of infrared blanching on enzyme activity and retention of β-Carotene and Vitamin C in dried mango. J Food Sci 80:E1235–E1242. Doi:10.1111/1750-3841.12866

Guida V, Ferrari G, Pataro G et al (2013) The effects of ohmic and conventional blanching on the nutritional, bioactive compounds and quality parameters of artichoke heads. LWT—Food Sci Technol 53:569–579. Doi:10.1016/j.lwt.2013.04.006

Icier F, Yildiz H, Baysal T (2006) Peroxidase inactivation and colour changes during ohmic blanching of pea puree. J Food Eng 74:424–429. Doi:10.1016/j.jfoodeng.2005.03.032

Jabbar S, Abid M, Hu B et al (2014a) Quality of carrot juice as influenced by blanching and sonication treatments. LWT—Food Sci Technol 55:16–21. Doi:10.1016/j.lwt.2013.09.007

Jabbar S, Abid M, Hu B et al (2014b) Influence of sonication and high hydrostatic pressure on the quality of carrot juice. Int J Food Sci Technol 49:2449–2457. Doi:10.1111/ijfs.12567

Jabbar S, Abid M, Wu T et al (2014c) Study on combined effects of blanching and sonication on different quality parameters of carrot juice. Int J Food Sci Nutr 65:28–33. Doi:10.3109/09637486.2013.836735

Jain P, Yadav DN, Rajput H, Bhatt DK (2013) Effect of pressure blanching on sensory and proximate composition of peanut milk. J Food Sci Technol 50:605–608. Doi:10.1007/s13197-011-0373-5

Knirsch MC, Alves dos Santos C, de Oliveira Martins, Soares Vicent AA, Vessoni Penna TC (2010) Ohmic heating—a review. Trends Food Sci Technol 21:436–441. Doi:10.1016/j.tifs.2010.06.003

Krishnamurthy K, Khurana HK, Soojin J et al (2008) Infrared heating in food processing: an overview. Compr Rev Food Sci Food Saf 7:2–13. Doi:10.1111/j.1541-4337.2007.00024.x

Latorre ME, Bonelli PR, Rojas AM, Gerschenson LN (2012) Microwave inactivation of red beet (*Beta vulgaris* L. var. conditiva) peroxidase and polyphenoloxidase and the effect of radiation on vegetable tissue quality. J Food Eng 109:676–684. Doi:10.1016/j.jfoodeng.2011.11.026

Lemmens L, Tibäck E, Svelander C et al (2009) Thermal pretreatments of carrot pieces using different heating techniques: effect on quality related aspects. Innov Food Sci Emerg Technol 10:522–529. Doi:10.1016/j.ifset.2009.05.004

Liu X, Gao Y, Peng X et al (2008) Inactivation of peroxidase and polyphenol oxidase in red beet (*Beta vulgaris* L.) extract with high pressure carbon dioxide. Innov Food Sci Emerg Technol 9:24–31. Doi:10.1016/j.ifset.2007.04.010

Manzocco L, Anese M, Nicoli MC (2008) Radiofrequency inactivation of oxidative food enzymes in model systems and apple derivatives. Food Res Int 41:1044–1049. Doi:10.1016/j.foodres.2008.07.020

Manzocco L, Quarta B, Dri A (2009) Polyphenoloxidase inactivation by light exposure in model systems and apple derivatives. Innov Food Sci Emerg Technol 10:506–511. Doi:10.1016/j.ifset.2009.02.004

Mukherjee S, Chattopadhyay PK (2007) Whirling bed blanching of potato cubes and its effects on product quality. J Food Eng 78:52–60. Doi:10.1016/j.jfoodeng.2005.09.001

Neves FIG, Vieira MC, Silva CLM (2012) Inactivation kinetics of peroxidase in zucchini (*Cucurbita pepo* L.) by heat and UV-C radiation. Innov Food Sci Emerg Technol 13:158–162. Doi:10.1016/j.ifset.2011.10.013

Palma-Orozco G, Sampedro JG, Ortiz-Moreno A, Nájera H (2012) In situ inactivation of polyphenol oxidase in mamey fruit (*Pouteria sapota*) by microwave treatment. J Food Sci 77: C359–C365. Doi:10.1111/j.1750-3841.2012.02632.x

Ranjan S, Dasgupta N, Walia N et al (2016) Microwave blanching: an emerging trend in food engineering and its effects on *Capsicum annuum* L. J Food Process Eng. Doi:10.1111/jfpe.12411

Rawson A, Tiwari BK, Tuohy MG et al (2011) Effect of ultrasound and blanching pretreatments on polyacetylene and carotenoid content of hot air and freeze dried carrot discs. Ultrason Sonochem 18:1172–1179. Doi:10.1016/j.ultsonch.2011.03.009

Schirack AV, Drake MA, Sanders TH, Sandeep KP (2006a) Characterization of aroma-active compounds in microwave blanched peanuts. J Food Sci. Doi:10.1111/j.1750-3841.2006.00173.x

Schirack AV, Sanders TH, Sandeep KP, Quality M (2006b) Effect of processing parameters on the temperature and moisture content of microwave-blanched peanuts. J Food Process Eng 30:225–240. Doi:10.1111/j.1745-4530.2007.00110.x

Severini C, Baiano A, De Pilli T et al (2005) Combined treatments of blanching and dehydration: study on potato cubes. J Food Eng 68:289–296. Doi:10.1016/j.jfoodeng.2004.05.045

Severini C, Giuliani R, De Filippis A (2016) Influence of different blanching methods on colour, ascorbic acid and phenolics content of broccoli. J Food Sci Technol. Doi:10.1007/s13197-015-1878-0

Sezer DB, Demirdöven A (2015) The effects of microwave blanching conditions on carrot slices: optimization and comparison. J Food Process Preserv 39:2188–2196. Doi:10.1111/jfpp.12463

Smith PG (2011) Introduction to food process engineering, 2nd edn. Springer, Heidelberg

Sotome I, Takenaka M, Koseki S et al (2009) Blanching of potato with superheated steam and hot water spray. LWT—Food Sci Technol 42:1035–1040. Doi:10.1016/j.lwt.2009.02.001

Stamatopoulos K, Katsoyannos E, Chatzilazarou A, Konteles SJ (2012) Improvement of oleuropein extractability by optimising steam blanching process as pre-treatment of olive leaf extraction via response surface methodology. Food Chem 133:344–351. Doi:10.1016/j.foodchem.2012.01.038

Viña SZ, Olivera DF, Marani CM et al (2007) Quality of brussels sprouts (*Brassica oleracea* L. gemmifera DC) as affected by blanching method. J Food Eng 80:218–225. Doi:10.1016/j.jfoodeng.2006.02.049

Xin Y, Zhang M, Yang H, Adhikari B (2013) Kinetics of argy wormwood (*Artemisia argyi*) leaf peroxidase and chlorophyll content changes due to thermal and thermosonication treatment. J Food Sci Technol 52:249–257. Doi:10.1007/s13197-013-0987-x

Printed in the United States
By Bookmasters